Such
Agreeable
Friends

SUCH AGREEABLE FRIENDS

Life with a Remarkable Group of Urban Squirrels

GRACE MARMOR SPRUCH

Drawings by Nurit Karlin

WILLIAM MORROW AND COMPANY, INC.
New York 1983

Library of Congress Cataloging in Publication Data

Spruch, Grace Marmor.
  Such agreeable friends.

  1. Gray squirrel. 2. Mammals—New York (State)—
Greenwich Village (N.Y.) I. Karlin, Nurit. II. Title.
QL737.R68S67   1983   599.32'32   83-8200
ISBN 0-688-02456-4

Printed in the United States of America

First Edition

1 2 3 4 5 6 7 8 9 10

BOOK DESIGN BY BERNARD SCHLEIFER

CONTENTS

Animals are such agreeable friends—they ask no questions, they pass no criticisms.

—GEORGE ELIOT

Such
Agreeable
Friends

1. IN THE BEGINNING

ON NOVEMBER 19, 1970, I got a birthday gift from on high. I was awakened at six A.M. by a crash. The steel thermometer I keep on the windowsill had fallen to the floor. The rubber plant near the window was bowing and swaying under some weight. I looked over toward Larry's side of the bed. He wasn't there. In the kitchen, no doubt, working. Where were my glasses? I squinted and made out a gray shape on a branch of the plant. My mind struggled to focus and make sense out of the swaying plant and the gray shape.

I have always loved squirrels. I love all animals, but squirrels are a special love, probably because of their shape—which makes the attraction physical, I suppose. This squirrel in my bedroom, on my birthday, was a clear challenge to the laws of probability.

I vaulted out of bed and into the living room, where I keep walnuts in a Papago Indian basket, and dashed back to the bedroom with the basket. The squirrel was on the dresser, waiting. I took a walnut out of the basket, but before I could offer it, the squirrel grabbed it from my hand and leaped out the window. I put the basket down and went to the kitchen to tell Larry why I was charging from room to room at that hour, but the sound of

11

a squirrel rummaging in the walnut basket made explanations redundant. Larry and I both got to the bedroom in time to see a tail disappearing out the window. We waited about ten minutes for a reappearance, then gave up and went back to bed.

I lay there marveling at the experience, but the experience wasn't over. During the next hour, fifty walnuts were transported, one by one, out of the bedroom and down the fire escape. At the end of that hour, Larry and I were ready to get up officially, according to the alarm clock. We hadn't been able to sleep anyway, for the traffic. I held out empty hands to the squirrel on the dresser, to indicate that there were no walnuts left. He stared at me, bit my thumb, got the idea after a while, and left.

The rest of the morning was taken up with telephoning the zoo, the Museum of Natural History, and the Health Department. The extremely sympathetic woman at the zoo, in answer to my question about whether squirrels can crack walnuts, asked if my visitor was a large squirrel. I thought he was, and she assured me he could handle walnuts. However, she recommended sunflower seeds, hard-boiled eggs, and peanut butter sandwiches on hard bread. She thought three or four nuts and a handful of sunflower seeds would make an adequate meal once a day, and that fifty walnuts was generous, to say the least. She reassured me that I had not killed my newfound friend with kindness, as he would not eat more than he needed and would store the rest for future repasts.

The Health Department assured me that there had not been a case of rabies among New York rodents in forty years (I winced at the word rodent), but recommended I have a doctor cauterize the bite if it had punctured the skin.

I telephoned the Institute of Animal Behavior at Rutgers University to inquire how I might establish a meaningful relationship with the squirrel. The mammal man I ultimately reached asked, "Meaningful for whom?" I learned that the best way to a squirrel's heart is through his stomach, and laid in a supply of sunflower seeds, peanuts, more walnuts, almonds, filberts, peanut butter, pumpernickel, and eggs.

At night, I put an assortment of nuts on the windowsill outside.

2. IN THE WILDS OF MANHATTAN

I LIVE IN MANHATTAN, on Eighth Street, Greenwich Village's Main Street, where the Fifth Avenue bus cuts across town at the end of its route. The house is vaguely Spanish-looking. It is a five-story walk-up, and I live on the top floor.

When we moved in, over thirty years ago, various members of Larry's family helped carry our belongings up the stairs and loaded our books on the dumbwaiter for us to pull up. It was our first permanent home, after a few months in a studio with rented furniture in Boston and nearly a year with my parents while we waited on a list for this apartment, owned by New York University. Correction: *leased* by New York University some years ago from Sailors' Snug Harbor for 252 years. Larry had just taken a teaching position in New York University's physics department, and that got us on the list.

We had told Larry's mother that we would be living one block from Washington Square Park. She could not help viewing a walk-up as a comedown. Having lived in one when she first arrived in this country as a girl, she knew that walk-ups were for immigrants, not young American assistant professors. Neither the eighteen-by-twenty-foot living room nor its seven-foot-wide

15

brick fireplace could compensate for the lack of an elevator—but Washington Square Park might. She thought she could see the park from our bedroom at the back of the house, not realizing the view would be blocked by houses bordering the park. She went to the window, gazed down at the communal garden, and as Larry and I chattered about Washington Square Park while carrying in boxes, she shook her head sadly and said slowly, "One tree."

The garden has more than one tree, in fact—if you consider the garden to comprise the space behind the entire row of houses rather than the space immediately behind our house. The garden is relevant to the appearance of the squirrel in our apartment.

The north side of Washington Square Park is lined with Greek Revival mansions built in the 1830's. Henry James's novel *Washington Square* is set in one of these houses, one that belonged to his grandmother and that was part of a group called the Rhinelander houses, which were torn down in the early 1950's to make way for a high-rise apartment building. Village pressure prevented the erection of the tall building directly on the square. Number 2 Fifth Avenue came in two parts, therefore; one towering and cream-colored, set back from the square, the other five-storied and red brick, a mock-Georgian structure on the spot where the five-story Rhinelander houses had stood. French proverbs notwithstanding, the more some things stay the same the more they change: Five stories in the 1950's were not five stories in the 1830's. The top of the new red-brick structure was several feet below the tops of its stately old neighbors. A false top was added to make it conform.

In the center of the north side of the park, at the base of Fifth Avenue, is Washington Arch, built by Stanford White in 1892. The northeast corner of the square, the corner nearest our house, still has the old buildings, at least outwardly. Numbers 7 to 13, just east of Fifth, were gutted years ago to make a single apartment house within the original façades. The remaining buildings house offices of New York University. Halfway between these houses and ours, a private way called Washington Mews runs parallel to Eighth Street. The Mews is a charming street, bordered mostly by two-story houses that had originally been stables. Our garden is between the row of houses on Eighth Street that includes ours and a row of Mews houses. If our squirrel lived in Washington Square Park, the path to our house would be: across the street to the old houses on the square, through their front gardens and around to the Mews, over the Mews' roofs to our garden, then up the ivy at the base of our house to the fire escape, and up the fire-escape stairs to our bedroom window. I say *"if"* because to this date I have not made a positive identification of any squirrel in the park—though one did do a double-take there on seeing Larry; nor have I found what might resemble a squirrel's abode in our garden, on our roof, or on any of the Mews roofs, but I do see squirrel traffic on different segments of the route just described.

In the mid-1960's, there was a major refurbishing of Washington Square Park after community pressure led to the banning of traffic from the park. (At one time, Fifth Avenue buses turned around in the center of the park.) The public was fenced out of some sections for months. During the refurbishing, the squirrels, with no dogs to harass them, may have multiplied. The park probably didn't have enough food to support the increase in population, and some enterprising squirrels fanned out into the neighborhood to forage. I remember seeing my first black squirrel in the park during the reconstruction. Ten years later, the park had dozens of black squirrels. One large tree appeared to house *only* black squirrels.

The morning after the squirrel's appearance, we were awakened at six-thirty by the sound of nuts being shelled. When the shelling was over, the sheller came in the open window, climbed up the rubber plant, down the rubber plant, up the tree lamp, down the tree lamp, leaped onto the dresser, where the walnuts had been the day before, and jumped onto the bed, trampling us in an obvious search for nuts. Unable, by nature, to refuse the requests of animals, I procured peanuts from a two-pound bag in the kitchen. I offered sunflower seeds as well, but the offer was declined. I had to leave for work by eight-thirty, so "relating" had to be cut short. I telephoned home from Newark before my first class at Rutgers University and learned from Larry that the squirrel had left shortly after I had, and had returned with a friend.

They, and friends of friends, have been coming essentially every day since, the word evidently having gotten out that, if you behaved, there was as much as you could eat at the Spruchs'.

Two days after the first visit was a Saturday. We had put peanuts outside Friday night to proclaim a welcome, lowering the window to prevent unwanted entry until a reasonable hour. But after the peanuts had been consumed (early), there was loud banging on the window—not on the glass but on the bottom of the sash, part of an attempt to dig into the apartment through the half-inch gap between the sash and the sill. We yielded, opened the window, produced more nuts, and resolved that this would not occur again. We would lower the window still further and would ignore any banging that might ensue.

Sunday brought success. We heard no sound before we opened the window at ten o'clock. We were probably so exhausted from premature awakenings the previous few mornings that we hadn't heard the peanuts being shelled outside. As soon as we opened the window, our friend came in. He was rewarded with a piece of hard-boiled egg—which he would have nothing to do with. Peanuts were substituted. Larry named him Chazzer, Yiddish for pig. Chazzer's friend showed up and stayed for a while. He or she seemed much shyer and less demanding. Chazzer was in and out all day, relieving us of about one pound of peanuts. As it got dark, he got frantic. Three peanuts remained on the sill and there was apparently not enough time for three trips downstairs to deposit them, and not enough internal room to contain them. When he started to bury one in the rubber plant, I put my foot down, and threw him out.

After about two weeks of visits, life with the squirrels developed a routine. At about nine each morning, the fire-escape window was raised to three inches above the sill. All but a section about seven inches wide, on the right, was blocked by several pieces of wood, which provided the squirrels with an entrance seven inches wide by three inches high and prevented us from freezing solid. Our bedroom is normally cold in winter, as we never turn the radiator on in that room. Now, "cold" became an understatement. Larry works at home many mornings, to avoid interruptions. As a theoretical physicist, the only equipment he needs is pencil and paper. He likes to sit near the fire-escape window, in the Danish black leather chair with teak arms. In order to survive in that spot when the squirrels' entrance was open, Larry would wear a sweater and a raincoat over his bathrobe and pajamas.

Chazzer's friend, called Little White Ears for the little white tufts behind his or her ears, was visiting more frequently than Chazzer. We wondered if the two were mates. We had not the slightest notion of their sex, but we had begun to wonder. We also wondered where they went at lunchtime since, even when their entrance was open all day, they would appear only in the morning and in the afternoon. We wondered whether they took siestas. We wondered about their social life.

One thing was clear. In only two weeks, they had been spoiled rotten. Six of the ten peanuts left out on the sill the previous night remained there in the morning. When we opened the window, a squirrel bounded in, leaped onto the bookcase, from there to the rubber plant, to the floor, and up onto the arm of Larry's chair. Larry thought he might be mad. I thought he wanted human companionship. Larry thought I, too, might be mad to have such an idea. I handed the squirrel a peanut. He took it and left. I said, "See?" When he returned, however, and I again held out a peanut, my hand was sniffed but the peanut wasn't taken. It occurred to me that another kind of nut was being sought. I got a walnut and held it out. That was accepted immediately. The walnut was being carried out of the apartment when, at the seven-by-three-inch exit, there was a "clunk!" Squirrel head plus walnut exceeded the exit's capacity. An adroit tilt of the head permitted departure with the treasure.

The squirrel returned, and in order to verify my interpretation, I initiated the previous sequence once more: my offering a peanut, its rejection, my offering a walnut, its seizure. Larry wanted to do his own verifying. He made the mistake, though, of having both peanut and walnut in the same hand. In the ensuing fracas, his finger was bitten. "He's completely crazy," Larry affirmed. "No," I insisted, "he's completely spoiled. We've shown him there are better things in life than peanuts."

3. LITTLE
WHITE EARS

I HAD SO MANY questions about the squirrels' behavior that I began to do some reading on squirrels and to make some inquiries. The reading wasn't very satisfactory. I learned that there are no less than seventy species of true squirrels distributed in all parts of the world except Australia and Madagascar, that our Eastern Gray squirrel, *sciurus carolinensis*, has a gestation period of forty-four days (maybe) and has been known to form large bands that migrate hundreds of miles, for reasons no one understands, crossing rivers on the way if need be. I found little on their day-to-day behavior. My inquiries produced only slightly better results. One expert said I could expect to see four or five males to one female, which would be larger and would stay at home. The woman I spoke to at the American Museum of Natural History said that, if anything, there were four or five females to each male, that both genders were the same size, and both were out foraging this time of the year—the female stays home with her young only in the spring. (My own observations were to confirm what the museum expert said.)

The scarcity of information and the disagreement among the experts made me decide to take notes on our visitors. As an experimental physicist, I am a trained observer—albeit of such personalityless subjects as electrons. However, if I could learn to refrain from anthropomorphic interpretation and simply stick to reporting observations, perhaps those observations might even be of some value. Our squirrels were in a category not considered in books: not pets, and not completely wild. They were city squirrels, living in many respects as squirrels do in the wild but depending upon humans for handouts. Ours went so far as to come into the house for their meals.

I held an internal interview with myself on my qualifications for writing a book on squirrels. My love of animals, one could argue, might interfere with objectivity. But many of those who study animal behavior are psychologists, and I have long been convinced that psychologists go into the field to solve their own problems—which would make them less objective and less able than I to observe without attributing motives.

I had been the mistress of fourteen turtles over a number of years, and I could boast having been bitten by, along with the standard animals, a horse, a swoose, and a camel. The first two were part of what is referred to these days as learning experiences. At age twelve, I offered a horse a sourball, at a Coney Island kiddie ride. Never having been instructed otherwise, I held the sourball in my fingers. The horse chomped down with his big horse teeth on sourball plus fingers. I learned the hard way to put tidbits on my palm when feeding a horse. Then he sweeps them in with his velvet upper lip. A lesson can be learned too well, however, or applied inappropriately. About ten years later, I applied it to a swoose, a half swan, half goose, like Alexander in the song popular at the time. The Philadelphia zoo had a swoose—named Alexander, in fact. I offered him a crackerjack on the palm of my hand. What is good for a horse is not good for the relative of a gander, I learned. Alexander's jaws closed over my extended fingers, missing the crackerjack. Although Alexander could not have been more than two feet tall, his jaws were an iron door slamming on my fingers.

The camel? I may not be able to claim an authentic bite. I was slowly moving back to take a picture in another zoo (I collect zoos) when I felt a blow on the back of my shoulder. Turning, I stared into the face, not two inches from mine, of a camel, a white one, whose body, many feet back of his face, was behind a fence. The face had given me the blow; specifically how—with teeth or forehead—I couldn't say. That zoo, incidentally, was in East Africa, and had been built for the benefit of local children, who could not afford to visit the game parks and would otherwise not see such animals.

Larry tries to top my bites with his having been peed on by a Barbary ape. On Gibraltar, we came across a man who made a livelihood out of getting some of the many apes there to pose with tourists for pictures. Larry despises picture taking, but was interested in the ape. The ape sat on his shoulder while I snapped the picture. Larry experienced a warm sensation—the ape's comment on the enterprise.

If I were to write about the squirrels, I would have to be able to identify them. I was beginning to suspect that one of our visitors was at least two. I thought of using spots of nail polish to mark them, like tennis balls, but Larry wouldn't hear of it. He was afraid a marked squirrel would be rejected by a mate or a parent or maybe even an entire commune. Nail polish on a squirrel did sound a bit like whoring after Babylon. The woman at the Museum of Natural History saw no problem in telling them apart. "Their faces are as individual as human beings' faces are," she told me. Larry asked, "Did you tell her you have trouble telling human beings apart?"

I phoned Rutgers' Institute of Animal Behavior again. As the institute had no squirrel expert, I was referred to the monkey man, Professor Hansen. He suggested a spot of dye for identification and assured me dye would not cause a squirrel to be rejected by his peers. Yet somehow I could not bring myself to do it. Putting dye on Chazzer or Little White Ears seemed a betrayal of trust. And they were indeed trusting. They would jump onto my lap and take nuts from my hand as I sat reading. They were getting tamer every day.

Meanwhile, Larry was getting fatter every day because of all the nuts in the house. Nuts are his, as well as squirrels', favorite food. By the end of the second week, he and the squirrels had consumed two and one-third bags of peanuts, each weighing one pound twelve ounces and costing eighty-nine cents, and three one-pound packages of mixed nuts at fifty-nine cents each. We were also stuck with a three-pound bag of sunflower seeds of which only a few had been eaten, during the first few visits, and peanut butter we didn't want in the house because of its abundance of calories and our lack of control. We were also out a few eggs. The peanut butter sandwiches and hard boiled eggs recommended by the zoo had been rejected by Chazzer and Little White Ears. The closest they came to peanut butter sandwiches was when I smeared peanut butter on a nut. It was licked off and apparently liked; but that had produced peanut butter pawprints in the bedroom, which *I* hadn't liked.

My friend Edie said the newspapers had predicted a hard winter because squirrels were busy burying nuts. The reporters had probably seen our squirrels burying the surplus nuts we had supplied. After a while, I developed enough sense to wonder what squirrels ate when they didn't have people like us to feed them; then I announced in my large lecture classes in general physics that I was in the market for acorns. That was good for thirty or forty pounds from a few animal-loving students who had access to oak trees.

My notes took the form of a diary. Each day I recorded what the squirrels had done that was of significance. The notes became, therefore, a record of what Larry and I learned about—and from

—squirrels. For example: It took a while, but eventually we learned that Little White Ears was not *a* squirrel but several squirrels. Gray squirrels grow tufts of white fur behind their ears in winter, to keep them from losing too much heat. Heat is radiated from the surface of an animal, or any object, for that matter. Ears are a large, exposed surface, relative to the volume of flesh they contain, and so a great deal of heat is radiated from them unless they are well insulated. The tufts are the insulation, and their whiteness reflects rather than radiates. The tufts disappear in the spring.

Animals in colder regions have smaller ears than their relatives in warmer climes. The polar bear's ears, relative to his size, are smaller than the brown bear's. Nature's design is to maximize efficiency. Larger ears may be better to hear with, my dear—useful to all, and particularly useful to prey—but they have the disadvantage of allowing more heat loss. The colder the climate, the more importance the disadvantage assumes, and the smaller the most efficient ear.

The size of the animal as a whole, however, goes the other way. Animals in colder areas are, in general, larger than their relatives in warmer places. A polar bear is larger than most other bears. If the animal is to conserve heat by radiating little, its surface-to-volume ratio must be low. That is, for a given volume, or size, the animal's surface must be as small as possible, since heat is radiated from the surface. Now, the surface-to-volume ratio is smaller for larger animals. To see this, consider a spherical animal—an idealization, to be sure, but a bear cub is pretty spherical if we overlook legs. Volume, being three-dimensional, is proportional to the cube of the radius of the animal. Surface, two-dimensional, is proportional to the square of the radius. The ratio of the surface to the volume, therefore, is inversely proportional to the radius—proportional to $1/R$, where R is the radius—and it gets smaller the larger the animal.

The surface-to-volume ratio also plays a role in metabolic rates (since there is little variation in body temperature from one species to another, and since small animals radiate more heat, they tend to have a higher metabolic rate per kilogram than larger animals) and in the ability to remain submerged (large aquatic mammals, such as whales, can stay submerged for longer periods than small ones).

Little White Ears knew nothing of surface-to-volume ratios, but was in any case admirably equipped.

4. EARLY EDUCATION, OR THERE'S NUTS IN THEM THERE KNEES

OUR EFFORTS TO KEEP the squirrels coming around, while not interfering in their private lives, were pretty clumsy, as I look back on them now. We fussed over them like ignorant new parents without the assistance of a Dr. Spock. But I must not anthropomorphize; I will strive for straightforward description.

Even observations are tempered by interpretation, however. What we notice or fail to notice is governed by our experience and our personality. My personality is such that I habitually construct elaborate theories from a single fact. Larry has exasperatedly called me "One Parameter Gracie." To head off difficulty from this source, I have made strenuous efforts to set things down in a way that should make obvious what is speculation and what is observation. Besides, a bit of bias may not be bad. Anthropomorphism may even be *in* this year. At one time, animals were thought of as being like machines, without feelings. Darwin provided some continuity between animals and humans, so that one could either "humanize brutes" or "brutalize humans." I've noticed, lately, that when I read about animal experiments in the newspapers, there is mention of such unmachinelike things as family relationships. A colleague of mine at Rutgers, Colin Beer, an animal behaviorist, said that at a recent conference he attended, words like "cost-benefit analysis" and "strategy" were used—this was a conference on ethology, not economics or poker. Today's metaphor apparently comes from economics and game theory. The metaphor changes with the times, and the different metaphors yield different results because they make us pose different questions. The way Beer put it was "the effect of observation on conception and the effect of conception on observation." It is a kind of extension of cross-fertilization between different fields. (I've often taken chauvinistic pride in the great many Nobel Prizes won by physicists in fields other than physics.) Some aspects of animal behavior have been interpreted using the mathematics of catastrophe theory.

The squirrel diary I kept started out with almost daily entries about everything, for everything was new and everything was interesting. The details that later ceased to be new did not cease to be interesting, but they formed part of the lore we acquired and were no longer recorded, except perfunctorily. What I now take to be common knowledge, had, in some instances, to be learned over several years. The diary became more condensed as time went on, and this book is a further condensation of the diary. I had intended to incorporate some diary entries directly. When I set about doing that, however, I realized that squirrel diaries are very much like human diaries: pretty boring except to the author and those who figure in them. Squirrels attend no important meetings and meet no fascinating people, not in our house, at any rate. The diary was simply a collection of bright doings, the analogue of a child's bright sayings.

I have tampered with chronology in order to present the biography of a squirrel as an entity. Yes, biography. The squirrels turned out to be as individual in character as the woman at the Museum of Natural History had said their faces were. Although I never learned to tell them apart from their faces—except for a few really distinctive ones like Scarface and Little One Eye—I could tell who was in the house from his or her behavior. It did not take long before these squirrels seemed like people to us, with characters, complexes, idiosyncrasies. We even developed likes and dislikes for different individuals. When I would come from work on a day I had left the house early, I would ask Larry, "Who was in today?" Over the years, there may have been more than one hundred squirrels visiting us, though no more than about half a dozen on any one day. There was even a genius among them, as we will see. I wondered whether, if the book were to be published, I could deduct nut costs from my income tax. Were the squirrels dependents?

In the beginning, we would put nuts outside on the windowsill before going to bed at night. We would close the window just enough to let air in and keep squirrels out. We were trying to train the squirrels to expect to be allowed in at a reasonable hour, so that if we wanted to sleep later than usual on a weekend we would be able to. Why we were putting nuts out the night before, when we were trying to train them to arrive after nine, is a mystery to me now. Maybe we didn't want to disappoint them.

One morning at about eight, I thought I would check whether the previous night's nuts had been taken. I peered through the inch of open window. My peer was met by a beady-eyed return peer. Another morning, Larry checked before nine; he lifted a slat in the blind and looked through. A squirrel sat on the railing of the fire escape, patiently waiting for opening time.

One night during this kindergarten period, it rained. Fearing the nuts would be all soggy and unappetizing by morning, I constructed a tunnel of aluminum foil to hold and protect them. I was awakened at dawn by the sound of aluminum being dragged across the fire escape. In my groggy state, I had a nightmare in which a squirrel got his head stuck in the narrow tunnel and, thrashing about blindly, neared the edge of the fire escape and fell off, smashing to the ground five floors below. The next night it rained again, and the nightmare still fresh, I made my second aluminum tunnel much wider. That awakened us long before dawn: Rain on an aluminum tunnel sounds like a steel band. Two nights of interrupted sleep finally produced the question that should have been asked in the first place: "So what if the nuts get wet?" As for the tunnels, there wasn't a sign of them in the morning. Not on the sill, not on the fire escape, not in the garden below. I suspected that if we were ever to discover our friends' abode, we would find aluminum mirrors on the walls.

At about the same time, the squirrels, tired of making the long trip downstairs to bury the nuts they weren't eating, began to bury them inside the house. I was witness to a burial in our large rubber plant before I was able to evict the burier, and noted that it was a very neat job: no soil on the floor, no visible evidence of burial. The soil that had been dug out for the hole had been gathered back in five or six wide gathering movements of each forepaw and tamped down. I dug where I had seen the hole made and came upon a filbert about two inches below the surface. The soil for the rubber plant is relatively soft; there are no roots until several inches down. The soil for that plant is now covered by chicken wire, painstakingly patterned to fit the pot and not interfere with the plant's growth, sharp edges taped to protect both plant and animals. Despite the chicken wire, on occasion a shoot pushes through the soil, twines around the trunk of the rubber plant, and bears no resemblance whatever to its host. It is an incipient almond tree.

The smaller plants were secondary burial sites. The nephthytis is very dense and full of surface roots. Burying is, therefore, not much below the surface. I caught a culprit in the act in that plant and yelled a loud "No!"—to which no attention was paid. I pushed gently. Complete absorption in digging. I pushed harder. Digging stopped. My hand was given two swift punches with closed paws. Digging was resumed. I put both hands on the digger's waist and tried to pull him off the job. Furious, he punched me repeatedly with both fists. The nut was still in his mouth awaiting interment—fortunately, I thought. I soon learned that a nut is never relinquished *under any circumstances.* I also learned that although squirrels use their teeth in skirmishes with one another, with me they would have only fistfights. Despite superiority in size, I did not win that battle over burial; the nut was laid to rest. Nor was this an isolated instance. Gentle, even timid, creatures become tigers when a nut is being cached away. The job done, the squirrel looked up at me for another nut. No hard feelings.

There were numerous other in-house burial sites: under the decorative pillows on the bed, on the bookcase under the drape, between the filing cabinet and the bookcase. I tried to unearth all the buried nuts and recycle them, but when Maude Davis, the woman who comes in to clean half a day each week, would do the bedroom, she would invariably turn up three or four I had missed. And every now and then there would be an "Ouch" from Larry when he put his foot in his shoe. Filbert in the toe.

It was interesting to watch an attempt at burial in the Navaho rug. The rug, a "Two Gray Hills," is predominantly gray, with a black and white pattern. Scratching—an attempt at digging—was done on the *black* parts. I think the black may have seemed like soil.

One Sunday morning, I was lounging in bed reading *The New York Times,* interrupted periodically by a squirrel jumping up onto the bed to be handed a filbert. The bedspread is dark green, making it resemble grass perhaps, for the squirrel would paw it a bit, discontentedly, and depart. Departing from a soft, flat bed was not accomplished by a graceful leap. At the edge of the bed, the squirrel squatted his rear end down, spread his hind legs wide, and plopped off onto the floor. A trot to the window, a graceful leap to the sill, and out the exit with the nut. The entire procedure was repeated several times. Finally, with pawing near the top of the bedspread, the nut went under. Satisfaction. No, a change of mind. To retrieve the nut, he began to dig. He could feel the nut between his paws but couldn't extract it from its green cover. A few more seconds of pawing and he placed the green-covered filbert in his mouth and began to gnaw. I yanked the bedspread from those needle-like teeth. He yanked back. I used both hands to pull. In the struggle, the nut came free. It was grabbed and carried off.

Probably the most interesting burial site of all was Larry's knee. Larry sat in his armchair near the window, reading, one leg crossed over the other. A squirrel, after searching for a suitable spot in which to leave his nut, and not finding any that met specifications, leaped up onto Larry's lap, made for the crossed knees, and proceeded to dig under the upper one. After the digging, the nut was pressed in between the knees, the surrounding area gathered in with the wide movements used for soil, and the spot tamped down. Somewhat more effort was spent on the gathering in and tamping down than was customary, this being the only indication that the terrain was at all unusual.

The episode indicated to me that squirrels do not view certain of our extremities as parts of us. Our eyes and our hands are us. Squirrels look straight into our eyes when asking for a nut, sometimes having to lean way out, as when seated on a shoulder, to do so. And a sudden movement of a hand can cause a squirrel to bolt out the window. Squirrels may also not view *their* extremities— their tails, to be specific—as belonging to them. Many a fellow has sat outside the exit with his tail inside our apartment; the window sash, three inches above, marking a dangerous divide.

The collar of Larry's bathrobe at the back of his neck was another attractive hiding place for nuts. Bathrobes in general appeared to be attractive, for a nut was buried under my bathrobe, with me in it, as I sat on the bed; the final tamp-down pat was on my rear.

The burying started me thinking, and I had what I thought was a great idea. I would get the squirrels their own private burial plot. I bought a large, rectangular, green metal window box, which I filled with earth and placed on the outside sill. It was a complete bust. There was a rainstorm the first evening the box was in position. Half an inch of water remained above the soil next morning. When I saw a squirrel digging below the water, I joyfully assumed he was burying a nut. Instead, he was *looking for* a nut. Not finding one in the mire, he tracked into the house, dragging his mire behind him.

Several days later, I was witness to the box being used first as a bathroom, then, immediately afterward, as a drinking trough. Though repelled by such disregard for sanitation, I left the box on the sill for several weeks, hoping it would eventually be put to the use for which I had intended it. But one squirrel, Hairy, developed a particularly disgusting habit. He had deposited on my hand a green slime that I thought might be excrement. When the green slime appeared again, I investigated. Instead of jumping over the window box, as the others did and as he used to do, Hairy had taken to walking through the mucky water before entering the house. I decided to get rid of the box when some of the water had evaporated; otherwise, getting rid of the green slime would involve a major cleanup. But the water above the soil ceased to evaporate as it became more covered with algae, and the box began to stink. I got rid of it. I spare you further details.

The plastic cup for the squirrels' nuts had more success than the window box. I would fill the cup with nuts and place it on the end of the inside windowsill furthest from the squirrels' entrance. The first time a squirrel used it, he knocked it to the floor, stared at it as it rolled slowly along, and then jumped down from the sill to watch from the orchestra. When I approached to pick up the nuts, he grabbed a filbert and ran. I replaced the cup. Essentially the same performance was repeated twice. A few minutes after the second encore, I looked up to see a squirrel standing silently on the sill next to the cup. The problem was that the cup, filberts buried below, appeared to be full of peanuts. I poked through the peanuts for a filbert. That was all the demonstration that was necessary. Next trip in, the squirrel did the poking. The time after that, the cup got dangerously close to the edge of the sill. One more visit and it would go over. No, it was steadied by a forepaw. The visit after that, however, the cup did go over, but not without a struggle. Forepaws were wrapped around the cup in a desperate attempt to keep it from falling. Only hindquarters were on the sill; the rest was going down with the cup. Crash! Confusion. Nuts all over the floor, the squirrel wandering from nut to nut. A terrible choice. I pointed to a filbert. That did it. My pointing had forced a decision. Out he went with his filbert.

The squirrels all ate outside on the fire escape at this time, an unsociable practice that the housekeeper in me nevertheless welcomed. Filbert finished, the animal returned and almost missed the cup, which I had replaced on the sill; he appeared to be heading for the spot on the floor where it had last been seen. Changing direction, he reached the cup and thrust his snout beneath the peanuts to reach a filbert, but was unsuccessful. He took peanuts out of the cup one by one, therefore, and laid them on the sill next to the cup. No filberts. He stared at me, willing me to get up from my seat and get some from the kitchen.

None of the fresh fruit the zoo had recommended had been eaten. Melon and apple had been turned down. A grape had been sniffed and rejected. When it was left outside at night along with nuts, everything was gone the next morning, including peanuts—which, increasingly, were remaining on the sill next morning, untouched—but the grape remained. There was also increasing fussiness about almonds. Since most of the nuts were being buried,

it appeared that the harder the shell the more suitable—less per-
ishable—the nut. Peanuts were still sometimes all right for on-the-
spot eating. When one was rejected, however—either by being
thrown violently to the floor, batted out of my hand, or simply not
accepted—the rejector would wait in the bedroom while I went
to the kitchen for something more savory. With almonds and
peanuts becoming more unappetizing, we wondered what new
delicacies we would have to come up with in order to provide our
charges with a varied diet. We did not want to be the source of
any deficiencies. We *hoped* that nature had a way of making
animals seek and accept what they needed.

I began to wonder whether I wasn't running the squirrels ragged
by offering them too many nuts to carry off and hide. Maybe they
are programmed genetically to squirrel away food for future hard
times and never to leave any to spoil or be taken by others. Or did
my playing Hermes to their Baucis and Philemon make no impression on them at all? Perhaps they noticed and simply considered it
their due. (As a child of the Great Depression—a factor that comes
right after genetic programming in producing compulsions—I
could appreciate their hoarding. When six rolls are put before me in
a restaurant, I have a great deal of difficulty leaving any. Fortunately, vanity and fear of obesity prevail. I look upon the doggie bag
as one of the great innovations of our time.) I had the disturbing
thought that there might be dozens of Chazzers and Little White
Ears. The woman at the museum had said "a handful of sunflower
seeds and three or four nuts once a day." A minimum of thirty-five
nuts were going each day. That would be ample for about eight
squirrels on museum rations. But I had seen Chazzer consume a
dozen filberts at a sitting.

Squirrels turn out to be remarkably clean. It was weeks before we noticed anything that might be associated with natural
functions. We had assumed that all such functions were being
performed on nature's premises. Then, either we became sharper-

eyed or our house became more like home, because we did find a few hard black pellets, ellipsoidally shaped, the major axis about eight millimeters, the minor axis about two. The hardness proved a desirable property, as it permitted easy picking up with paper toweling with not a trace left. Hardness seemed consistent with a diet of nuts. As for urinating, we at first thought that squirrels didn't require it, but we later became more knowledgeable. We were to learn that strained circumstances could call forth a few drops of urine, the squirrel equivalent, perhaps, of peeing in one's pants. (Apart from one possible bladder problem: An individual produced what could be termed a veritable torrent for his or her size, on a copy of A. P. French's *Special Relativity*.) If not wiped up immediately, I am convinced squirrel urine would eat through steel. It did, in fact, eat through, in only seconds, the linseed oil on an old English writing box, bleaching the mahogany. As these drops were few and far between, and what there were could be handled by contact paper on well-traveled surfaces and swiftness elsewhere, we never gave a thought to toilet training. (I don't know what we could have done if we *had* given it a thought.) There was a single act of what appeared to be spite and defiance: a deliberate squatting down on our bed, despite shouts and threatening gestures. The objects that suffered most were books. Many are marked with muddy pawprints, from rainy days or times when the rubber plant had just been watered and I was lacking in speed. But every print is dear.

5. WHAT'S IN A NAME?

Now THAT WE KNEW there were multiple Little White Ears, we set about trying to distinguish among members of this subset of squirrels and to assign them meaningful names. We knew we couldn't give them names like Charlie, and then have to say, "Charlie? He's the one with the scraggly tail and dark face . . ." We would have to name them directly after a distinguishing feature, however puerile that might sound. For the time being, we could avoid puerility by dubbing one Slim, another Hairy, for these were apt, single-word descriptions. We were to find that these names would not do either, for they corresponded to features with no permanence. But that was still to be learned.

We had made one important step in our education: We could tell the males from the females—when they sat up. The penis is quite visible when the male sits on his haunches, to eat a nut, for example. As squirrels weren't always sitting up in our presence, and standing up on their hind legs even less often, we did not always know which sex we were dealing with. The female's underside is a smooth expanse of white. It seems almost wrong to refer to it as an underside—it is a soft, furry, warm-looking, inviting belly.

37

In the same way that we were studying them, one of them, Slim, seemed to be studying Larry. The little fellow sat up on the arm of Larry's chair, head cocked, front paws turned in toward his chest, and stared into Larry's face for at least one full minute. He did this several days in a row, his periods of contemplation getting longer and longer. What was so fascinating about Larry? Was he trying to distinguish Larry from me? Larry began to feel uncomfortable. Was Slim thinking he was the biggest nut he'd ever seen? It may have started when Slim saw Larry on the floor doing back exercises. He couldn't get over that.

Slim was very relaxed in the house. After contemplating Larry, he would explore the bedroom for possible nut-burying sites. Under the black corduroy bed pillow seemed to be his favorite. He was indifferent to my removing the nuts, and so relaxed, or confident, or both, that he conducted his explorations while I moved about the room.

His explorations soon took him to a frontier, the open doorway of the bedroom, which led to the long hall, with the kitchen on the right, three feet from the bedroom door. He was very cautious about exploring the new territory, slowly stretching his back legs to their maximum length—they remained fixed as his front legs moved forward—as if trying to keep hold of familiar terrain as long as possible. (This would turn out to be characteristic behavior in the face of the unknown.) He went only as far as the entrance to the kitchen, this time, before turning back. Enough adventure for one day.

Back in the bedroom, he started to bury a filbert in the nephthytis plant, changed his mind, and tried to exit through the green plastic plant support. Slim as he was, he was too fat for the support. He was stuck for a few seconds. There was no panic, no fright, just a lot of pushing and squeezing.

Slim seemed an adventurous type. He did two other noteworthy things that day. First, we saw him rummaging with his front paws in the nut cup on the windowsill, then making for the exit rather clumsily. It was his clumsiness that attracted our attention. He had moved less than a foot toward the exit when we understood the cause. He was attempting to carry out two filberts at a time. One was in his mouth, the second was supported against the first in his front paws. He had only his back legs with which to walk. Although squirrels stand up on their back legs very well (usually with their front paws turned in at the wrist so they touch their chest) they walk no better on them than a dog does. The arrangement wasn't very satisfactory to Slim, so he expelled the filbert from his mouth and held both side by side in his paws. That wasn't good either, so back one went into his mouth. The filberts went from hand to mouth to hand in an amateurish exhibition of juggling. Finally, one fell to the sill. There he left it, hesitantly, and with what appeared to be regret. After a few steps toward the exit, there was a look back, and a second backward glance before departure. Juggling would become a fairly common sight, as each squirrel had to try his hand at carrying two filberts several times before learning it couldn't be done. The juggling act even served to indicate a newcomer on the premises.

Slim's other innovation was a leap onto Larry's shoulder. Startled, Larry must have recoiled, for Slim leaped right off again. Next time, Larry was prepared. He didn't move, and Slim remained there to eat his filbert, sending shell fragments into Larry's ear.

Prior to our intimacy with squirrels, we had not stopped to wonder how a one-pound squirrel could open a filbert when a one-hundred-fifty-pound human needs a nutcracker. The secret lies in pressure—used in the technical sense—and needle-sharp teeth. The technical definition of pressure is force per unit area. A petite one-hundred-pound woman can exert more pressure in stepping on your toe than a two-hundred-fifty-pound man if, say, both step on your toe with all their weight on one heel. The man's heel is about three inches by three inches, which gives an area of nine or ten square inches. If we divide the force he exerts on your foot, his weight of two hundred fifty pounds, by the area of his heel, ten square inches, we get a pressure of twenty-five pounds per square inch. If the woman wears shoes with narrow heels, say about half a square inch in area, then her weight of one hundred pounds on an area of half a square inch produces a pressure of two hundred pounds per square inch, eight times that of the two-hundred-fifty-pound man. If a squirrel can exert a pound or so of force with his jaw, and that force is applied to a nut by a tooth about .001 square inch in area, the pressure that results is one thousand pounds per square inch. (If we're wrong in our estimate of how much force our squirrel can exert, and he can actually exert only half a pound instead of one pound of force, the pressure will be half as much, but still five hundred pounds per square inch.) Even with that much pressure, however, the squirrel prepares his nut. A filbert is first rotated a few times in the forepaws and got into position. Then sawing starts, usually on one spot. Next, the nut is pulled out to the side, a tooth applied to the sawed region, and pressure exerted to crack it open. A tough nut requires sawing in more than one spot. The entire process takes about ten seconds for an average squirrel, three seconds for a champion, and more than half an hour for a game but sickly little fellow named Runty, about whom we will say more later. The very young can't manage filberts at all.

It was interesting to us to see that there were two classes of shells, with all the shells in one class alike. There seemed to be two styles of shelling. In the more common method, after having been sawed as if by a rotary saw, about one third of the shell is cracked off, like a cover being removed. The remaining two thirds are held in the paws like a plate, to hold the meat while it is being eaten. If the meat cannot be removed completely, there is more cracking, but no more sawing. When the meat has been consumed, the plate-shell is dropped. In the less common style, the shell is split in half. Work on a peanut is trivial. It will be bitten at one end, turned around, bitten at the other end, turned again until the proper end is selected and actual labor begins. One or two bites into the shell and the meat is exposed and pulled out. Sometimes the second half is laid aside to be taken up again when the first has been eaten. At other times, the second half is used as the plate for the first half.

We noticed that, every now and then, a filbert would be taken from the cup, rotated to get it into shelling position, then dropped on the floor. Although we at first attributed this to clumsiness, we soon realized the nuts weren't being dropped, but *thrown* to the floor. We investigated. We examined a thrown filbert. It looked like any other filbert *on the outside.* One shouldn't judge a filbert by its cover, we learned. Cracking showed it was empty!

We inspected more of these rejects. All empty or with a dried remnant inside. How do squirrels know in advance not to waste sawing effort on an empty nut? It could not be smell; there was no sniffing before the nut was thrown away. But there was rotating! The rotating that we had interpreted as getting the nut into sawing position was either not that, or that and something more: It indicates whether there is a nutmeat inside. To check this hypothesis, we shook several filberts, then cracked them open. By George— or by Slim—we, too, could tell! There is rattling when there is a full-fledged nutmeat inside. The squirrel's rotating served, in physics terminology, to measure qualitatively the nut's moment of inertia. A one-pound expert on nuts. I might add that now I reflexively shake a filbert before applying a nutcracker.

Shelling nuts serves a purpose over and above that of reaching food. Squirrels' teeth do not stop growing at a certain length, as humans' do. Humans can get by with limited-length teeth because gnawing isn't their business, as it is for rodents. (Rodent means "gnawing animal".) Nonstop tooth growth is characteristic of rodents. Their teeth must be filed down or they will penetrate the brain case. Honing top teeth against bottom ones is one means of filing. Apparently, this is not entirely satisfactory; we hear of squirrels gnawing telephone wires and presume it is not the conversations inside that attract—or disturb—the gnawers.

Our rotting window frame appeared to be good filing material, or perhaps the bits of wood on the outside sill constituted evidence of attempted break-ins while we were out. At any rate, our fire-escape window was becoming a casualty of friendship, which I did not think our superintendent, Mr. Lassiter, would appreciate. I tried to prolong its life with a foot-high fence of chicken wire on the outside windowsill except for the small entrance space, and was in the act of installing it when one of my friends brushed unperturbedly by hammer and hand to push on to the cup of nuts. It was beginning to appear as though the squirrels thought *they* lived in the apartment.

6. STILL LEARNING

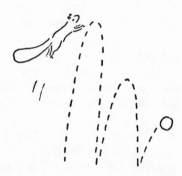

ONE DAY AFTER the official start of winter, there was an inch of snow on the fire escape to mark the date. We opened up shop at the usual hour of nine, but there were no customers, not for hours. This was our first experience of what would become a common fact: Squirrels don't come out in very bad weather. If they're caught in the rain, they will make the best of a bad situation, particularly if they're in the middle of lunch. But if it is extremely cold, or raining hard, or snowing, they will wait. We envisioned them curled up, taking late-morning snoozes.

Toward noon, a little fellow showed up. He wasn't one of our regulars; he was quite shy and hesitant about coming in. But he knew about the cup of nuts, either by smell or from having followed one of the regulars on an earlier occasion. Faced with the treasure all to himself, he went wild. Fourteen ounces of greed. He charged in and out, burying the entire contents of the cup in the snow on the fire escape in a few minutes; there was a neat almond border at the base of the railing. Then he raced away on the top of the fire-escape railing. I gasped. The railing was slippery with ice. He had flopped off the flat, nonslippery windowsill onto the floor of the bedroom. Larry and I thought about it and concluded that the shy newcomer was not an oaf of a squirrel; *all* our visitors had difficulty on smooth, flat surfaces. They slithered around on the dresser or the wood floor, their legs sometimes slipping apart under them in a double split. The railing of the fire escape could be grasped, ice notwithstanding; when they were able to grasp, they could get around without difficulty. Their forepaws are like hands, with long fingers that they use much as we use ours. The hind feet also have long fingers, which are used for climbing or hanging face down, but they don't seem to be used for manipulative operations like shelling nuts.

I got a good view of the hanging procedure once when a squirrel buried a nut between the mattress and the box spring of our bed. He hung by his back paws from the edge of the bed as from a trapeze, while his front paws did the burying eight inches below. His back legs were spread wide and turned out like a ballerina's, his paws turned out even further, so that they were 180 degrees with respect to their walking position, namely backward, with the palms (or soles) against the bed and the claws dug in. If you were to turn your wrists so that your palms face forward as your arms hang down, then take your arms straight back as you bend forward and hang from your hands, that would correspond to the position of the squirrel's back paws.

Claws are arranged well for going up, not down. It's easier for a squirrel to go down, though, than for a cat, which doesn't turn its back paws around to hang. That is why cats are frequently rescued from trees they were able to climb up. We don't often see the ASPCA or firemen on ladders rescuing squirrels from trees.

In jumping, too, up is handled differently from down. We realized this after many a missed heartbeat when a squirrel would leap from the floor of our fire escape to the railing. Missing the railing meant a plunge of five stories. But they never missed. And the reason they never missed is that they have plenty of time to grasp the railing when at the top of a leap. The situation is quite different for jumping down. In fact, we started thinking about the problem when we noticed that the squirrels leaped up to the railing from the floor of the fire escape but would climb down one of the vertical bars. Conditions are the same as for a ball that is thrown. When a ball is thrown upward, it is motionless for an instant at the top of its rise just before it starts down. That is, when one throws a ball upward, the ball travels most rapidly the instant it leaves the hand, travels more slowly the higher it gets, being slowed by the force of gravity, until it is motionless at the top of its path. Starting down, the ball accelerates and is moving fastest at the bottom of its path. In the same way, a squirrel jumping down experiences his greatest speed just before landing; jumping up, he experiences his slowest speed just before landing, so there is plenty of time to grasp the railing. We stopped worrying.

Learning was not one-sided. The squirrels were gradually learning about us and our possessions. During their first few visits, they would climb everything climbable: the tree lamp, the rubber plant, the burglary-prevention grill. They stopped after a while, having learned which were the dead ends.

The ring of the telephone sent them flying at first, before they became accustomed to the sound. But even during the period when the phone bell could cause panic, Larry's sneeze, the loudest I know—it makes walls ring, and has frightened many a human —only made Slim withdraw from the arm of the chair to the windowsill to continue munching his filbert. Either a sneeze is recognized as an unfrightening human sound, or it takes more to interrupt a meal. Sounds that *do* interrupt a meal include not-too-distant sirens—ambulance or fire engine—a dog's bark, voices in the garden. Then there will be a rush to the edge of the fire escape, a lean far over the edge, powerful little shoulder muscles bulging, to see what the disturbance is all about, an unfinished nut protruding from the mouth the entire time.

Self-induced sounds, such as that produced when the cup is

knocked off the sill, are heard with equanimity. Squirrel-made sounds are recognized: One individual was "burying" in the middle of the small barrel we use as a wastepaper basket—in the middle because, the bottom being round, no corner could be found—and the tamping down was amplified to loud thumps, causing an associate on the way to the cup to stop, hearken—head cocked—until the thumping stopped. A gong on television caused one squirrel to sit bolt upright, while the sounds of war—shooting, bombs exploding—were apparently not of consequence. As Larry summed up their concern with sounds: "Noises are their business."

Larry tried to teach Slim about pockets. He had put filberts in his bathrobe pocket and sat reading in his armchair. Slim evidently smelled the nuts. With a bit of guidance, Slim put his head into the pocket to take one. More and more of him went into the pocket until only a round little rear stuck out. After removing himself and eating his filbert, he returned for another, but this time he felt the nuts before he got to the pocket opening. It was clear that he didn't remember how to get into a pocket. He became frantic and began to gnaw the fabric. Larry had to take a nut out and hand it to him to save the pocket.

Ever inclined toward athletics, Larry insisted on testing and developing the squirrels' strength. He would hold onto a nut to see how hard it would be pulled from his hand. He was impressed. One might think that a creature weighing one pound could not exert more than one pound of force, but that is not the case. Little children often lift others weighing more than they do, and certainly professional weightlifters do. We have all experienced pushing or pulling with a force greater than our own weight, provided that we have something to press against.

It was the squirrels' phenomenal athletic ability—their incredible agility—that caused Larry's interest in them to increase. He was becoming attached to them. He had had cats and kittens throughout his childhood, but had never been close to any other animal. He may even have seen squirrels from a cat's-eye view. I had not been very close to squirrels either, but when it is *appearance* that attracts, one can love from afar. One might say the same about admiring agility. But we would not have noticed the agility unless we had been previously attracted through appearance. For whatever reasons, squirrels were assuming more and more importance in our lives.

7. SQUIRRELS' LIB

ONE MORNING I HAD BEEN feeding Chazzer, whom I identified by his large size, somewhat yellowish cast, sunken cheeks and nervous, jerky movements. Not at all graceful, he sort of galumphed in. He took his nuts outside, to eat on the fire-escape railing, and was on his fifteenth filbert when suddenly he began to half-bark, half-scream raspingly, his entire body heaving with the effort. I thought he was vomiting. He had gorged himself. But no, he wasn't sick; another squirrel had come into view and Chazzer was barking at him.

I had not realized that squirrels made sounds, and if I thought about the subject at all, I suppose I had assumed squirrels were silent, like giraffes. (It turns out giraffes aren't silent either; they just don't use their low, fluttering voices very often.) But here was Chazzer barking, sounding somewhat like a duck quacking. Additional squirrel sounds would soon impinge upon my consciousness: gurgling, gibbering, carping, a bok bok bok and a growl. Can you imagine a gentle little squirrel growling? Well, I have been growled at by squirrels, and I also now know that not all squirrels are gentle. Anger, and anxiety, too, I think, is expressed by clicking the teeth. Excitement, or agitation, has no sound; the tail flicks up and down (in circles for some individuals).

Chazzer's barking did not have the quality of casual conversation. He was *yelling* at the other squirrel, like a boss giving an underling a dressing down.

The idea that there might be a kind of hierarchy was taking form in my mind. A few days later, the fellow that had been in Larry's pocket showed up. I thought it was he because he headed straight for Larry's pocket, but he might have been following his nose, of course. He sniffed Larry's fingers, pushed Larry's hand aside with nose and paws, extracted a nut, and selected the bookcase as the spot for his brunch. He'd eaten a few nuts when a squirrel's head poked in the entrance. No motion on anyone's part for several seconds. Then the newcomer pulled back and brunch was resumed. But the newcomer returned, dared to enter. A lunge drove him out. Several times there was an in, a lunge, an out, until I couldn't stand any more and smuggled a nut outside to the have-not. The nut provided the courage and/or the strength to get in and stay in—at a distance great enough from the first, on the dresser, to allow retreat if necessary. The two maintained an uneasy peace through several filberts on each side when zyn! the newcomer stepped on the minefield of the typewriter. Retreat in panic.

Age appeared to be deferred to, and that seemed reasonable. How could I tell age? In much the same way one can tell the age of a dog. A young dog is usually slimmer, firmer-looking than an older dog. Squirrels' fur is not very figure-revealing, particularly in winter, making it difficult to tell age from shape. To my great astonishment, I could tell from their faces. After several months of visits, I had begun to notice smooth faces, eyes big in the head and sometimes outlined in a lighter color, giving a Bambi-like appearance, and a facial fuzziness on some, all of which appeared youthful to me, while others had hollow cheeks, sometimes with a yellowish cast, and bags under their eyes. The latter squirrels moved more slowly and were usually less frisky than the smooth-faced ones. I assumed they were older. There was a difference in size, but not much. The youngest visitors appeared to be two-thirds the size of the older ones.

Size is related to one aspect of the hierarchy that bothered me no end. Females appeared to be the same size as males. That is, mature females were the same size as mature males, as far as I could tell. *Yet females were lower in the hierarchy.* An adult female might be higher up than an adolescent male, but when male and female were about the same age, the male was boss. If a male was present, a female could not assert herself directly. She was always chased from the cup, and had to resort to wiles and maneuvers to sneak nuts out. We would witness this repeatedly, with pairs, with several I believed to be brother and sister, and with non-relatives. And the anthropomorphizing women's libber in me resented it. *With no difference between males and females that I could see, the females were second-class citizens.*

One evening, I met Virginia Tiger, a colleague at Rutgers, in a restaurant in the Village. Virginia was there with her husband, Lionel Tiger, the anthropologist, who had written a book on animals—and humans—with Robin Fox. When Virginia introduced me to Lionel, I launched into an exasperated query on why female squirrels were number two. Well, first we had to establish that I was using "number two" in a very restricted sense, namely with regard to food, and not, say, life-span, because in some species females live longer. Then, after ruling out birds (many male birds feed their females), and stating that each species should really be examined individually, Lionel acknowledged that for some small mammals, such as squirrels, the male had fewer ongoing responsibilities and, therefore, if the female wanted—or needed—to keep him hanging around the den or nest, she had better defer to him.

Now, I wouldn't exactly call human women small mammals, but what Lionel said jibed with an article on human male-female relations that had impressed me greatly, written by Diana Trilling. Trilling had stated that, since women appeared to be in a secondary position in virtually every type of society one could mention —capitalist, communist, kibbutz, etc.—that was evidence that the secondary position was not culturally determined. Her conclusion was that women needed men for sexual or maternal satisfaction, and that if men's sexual performance depended upon their feeling that they were in the top position, women were prepared, consciously or unconsciously, to accommodate.

Somehow though, in the case of squirrels, accommodation seemed crass. No subtlety, no pretense; it was right out in front. Females had to wait their turn at the cup. I felt sorry for them. They had no options. Compassion would change my dislike of a nasty little female named Halftail to tolerance. She would lunge at my hand with a growl if my hand went near her while she was eating, even if that hand was putting more nuts down. She gave me quite a few scratches with her nails. Sometimes, if I was a bit slow in opening the window or the screen, I would get a growl, always a lunge. I could never hand her a nut. I could not even put it down. I had to throw it down, or be scratched. I came to realize, eventually, that there was a reason for her nastiness. She simply did not have the time for me to put a nut down slowly. She had to consume that nut before she was chased from the premises by some male. She was the fastest eater in the East. She had to be, poor thing.

8. RUNTY

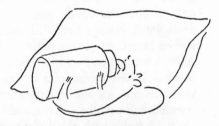

THE SQUIRRELS CAME to us in batches. Some of the three or four—at most half a dozen—that would appear on any one day would become regulars and would continue to appear for months, for years in a few cases. With some overlap in time, one batch would be replaced by another. People have asked me what the life-span of a squirrel is. Some sources say five or six years in the wild. We had two squirrels visit us for that length of time before they disappeared. I read of a squirrel kept as a pet who lived for twenty years, but I *know* of a pet squirrel found as a baby and kept for six years, at which time it turned gray, then white, went blind, and died. The life-span in the wild is not a "true" life-span, for when animals get older and slower, they find it harder to compete for food and easier to become prey. Whether the slowness of elderly Washington Square squirrels renders them prey for dogs and cars remains a question for me, for I believe that the increase in age brings a concomitant increase in skill in avoiding dogs and cars. In the only instance I know in which a squirrel was caught by dogs in the square, the squirrel was very young. Five dogs had cornered it. One had seized it in his teeth and evidently broken the little creature's back; it was dragging its hind section. Larry came upon the scene just after the startled dogs had been called off. (I believe dogs don't expect to catch squirrels; the game is in the chase. Catching a squirrel spoils the game, makes it serious. City dogs don't know what to do with a squirrel once it is caught.) A policeman refused to shoot the squirrel; it was too small, he said—he has to shoot from five feet away. He offered to club it, instead. The offer was declined. Two New York University students got a shoebox from their dormitory nearby and put the squirrel in it. Larry and one of the students took the squirrel to the ASPCA hospital on Ninety-second Street. The examining doctor confirmed the diagnosis of a broken back and said the squirrel would never walk again. He recommended it be put to sleep. Larry assented, and paid the very reasonable five-dollar fee for the visit. On the bus ride back, the student confessed that he would have wanted to take the crippled little beast home with him.

Mrs. Russell, our remarkable neighbor on the third floor who is now past ninety-six (I think she has taken to lying about her age a bit, adding a year when she wants to impress people), informed me that there were four squirrels in the Eighth Street gang. She, too, was feeding them. When a meal was ready, she would call them with a long, loud, kissing sound, to which they responded instantly, she asserted, if they were anywhere in the garden. She served shelled pecans on her outside windowsill. Mrs. Russell and I would both be in trouble with Superintendent Lassiter soon, but Mrs. Russell was tougher than I and could handle him better.

This first batch of squirrels included Slim and Hairy and Chazzer and a few transients like Little One Eye, who had appeared on a January day when the temperature was 12 degrees Fahrenheit. He was given his name because one of his eyes was more than half-closed. Probably, he was unable to see with that eye, for when he jumped onto the arm of Larry's chair, he didn't quite make it and flopped off onto Larry's lap. Larry helped him back onto the arm, from which he leaped onto the bed, landed facing away from Larry, turned around to accept the nut held out to him, jumped to the floor, went over to the rubber plant and up onto the top of its wooden pot, then ricocheted off the left branch onto the sill and left. Little One Eye returned several times for more nuts, traversing the same complicated path each time, including falling off the arm of the chair. If it had to do with trying to improve his field of vision, we couldn't see how.

Little One Eye visited us only a few times. On his second visit, he stopped on the bed in mid-ritual when he spotted the television set. He stood and watched for about half a minute. The Superbowl game was on.

Naturally, some squirrels made more of an impression on us than others. The first one to make a deep impression arrived one afternoon in the middle of March, near five o'clock, as I was sitting in the armchair near the window, grading papers. I saw a tiny head appear in the squirrels' entrance. The head was all I saw; it would come in no further. I held out a filbert. It was accepted after my thumb was bitten very gently. An inordinate length of time was spent on that filbert. I stood up and studied the new fellow through the window as he ate on the outer sill. He was much smaller than the others, perhaps half their size, and had a scraggly tail.

I put the cup of nuts closer to the entrance, sat motionless in my chair, and waited ages for him to finish the filbert and come in for more. He finally entered, appeared not to notice me, went to the cup, and took a filbert. After half a minute's effort to crack it, he dropped it on the floor, deliberately, and took a peanut, which he shelled and ate fairly rapidly. Half of each nut in the double shell was dropped on the floor, not deliberately. He took another peanut from the cup. After a few seconds, *it* fell to the floor. This time he looked at it with (what seemed to me) regret. It was a big jump to the floor. He had not come in far enough to see the novice trail through the rubber plant. There seemed to be indecision as well as regret in his movements. Should he go after it or shouldn't he? He made up his mind; he took a peanut from the cup. I was anxious for him to leave, as it was getting dark. He was in the house much later than any of the others had been. Since the squirrels all left before dark, I wondered if they could see well enough to get down the fire escape after sunset. I assumed they went to bed early. Perhaps the tiny fellow didn't know how dark it was outside; the light was on in the room. Just then, he noticed me; I may have moved. He gave a little start. I sort of cooed at him, assuring him I was friendly. He went on eating. I was considering evicting him for his own good when he took a peanut and left with it. The floor was littered with shells.

During the following week, the tiny fellow came back a few times, carefully investigating the premises each time. As soon as he came through the entrance, he would stop to look down at the floor, then up at the leaves of the rubber plant. Only after eating a few nuts would he focus on Larry or me. He ate on the window-sill inside, peanuts and filberts; the filberts took him forever to crack.

Over the weekend, Larry and I had gone to a wedding. The bride was Jewish and the groom Protestant, and so they had had one of those offbeat weddings popular at the end of the sixties: This one was Hindu, performed by an authentic guru. At the reception afterward—where I was shocked to see the guru devouring meat hors d'oeuvre—I met the groom's uncle Douglas, who came from a small town in Wisconsin and knew all about squirrels. Uncle Douglas told me the tiny fellow was, very likely, a runt rather than a youngster, because a youngster would have a bushy tail. He said youngsters are tended for about twenty-five to thirty-five days, after which they are almost the same size as their parents and are on their own.

The runt idea made sense. Our tiny fellow seemed so frightened; probably, everybody picked on him. Runty was inside when another squirrel with a scraggly tail showed up. Runty was terrified. He dropped to the floor, his little frame heaving with heavy breathing. He waited there, transfixed, until the other one left; then he leaped out. As a matter of fact, I had thought Runty was blossoming on our nuts, until I realized there was another with a tail like his.

Runty had a yellowish back foot, and a slightly yellowish cast to his haggard-looking face. He also had what appeared to be an indentation in one ear, which I was not sure he had been born with —it looked like the result of a bite. He had the white chest all the gray squirrels had, but he had a disturbing bulge on his belly. I wondered if it was a tumor, or the bloated belly associated with starvation in humans.

We began to fuss over him. Larry cracked a filbert for him. I left a peanut in the entrance in case he was too frightened to come in. All was in vain: Slim marched in with Runty's peanut in his mouth, to eat on the bookcase near me.

We had gradually switched the overnight supply of nuts to peanuts, as being quieter than filberts in dawn consumption. We figured that the spoiled squirrels wouldn't eat them, knowing filberts awaited them when we awoke, and the poor souls like Runty would always get something, if they wanted to play early bird. However, Larry looked out of the window early one morning and saw that the early bird was, in fact, a bird: A blue jay was flying off with a peanut! For weeks, two blue jays had been eyeing squirrels emerging from our window with booty. Then, one morning, seized by resentment or envy or the simple desire to get the nuts, they attacked! They dive-bombed Slim, who sat on the triangular corner of the fire-escape railing, minding his own nut, provoking no one. Slim ducked. A jay turned and swooped down again. And again. Slim ducked each time. Between bombings, he went on eating, never moving from his spot on the railing. All over the country, people were trying to keep squirrels away from the food they put out for the birds. Now we would be trying to keep birds away from the food we put out for the squirrels.

Two days later, at five-forty in the morning, I heard the crunching of peanut shells. I dragged myself out of bed and saw Runty. He had a characteristic hiccup. (It may have been a twitch, but it looked like a hiccup.) As I had expected, he was finding it necessary to come at that hour to avoid the others and assure himself a meal.

The dawn crunching of peanuts continued through the month of April, usually awakening us, but because we believed it to be Runty, we didn't mind and fell asleep again. One day, Runty showed up in the afternoon, when we were home, and took a filbert. He sucked it and sucked it, for over half an hour, finally managing to make a small hole in the shell through which he somehow got out the meat in bits. Larry and I couldn't bear a half-hour of sucking for each nut, which made a noise loud enough for us to be unable to work. Larry cracked some filberts with a nutcracker and left them on the sill. (Runty preferred not to take nuts from our hands.) I decided to try an experiment. I poured some milk into a small aluminum container and placed it on the sill outside. Runty sniffed at the container, then jumped up onto the fire-escape railing, smacking his lips. He climbed down, investigated the milk again, but took a peanut. After eating the peanut, Runty returned to the container once more, stuck his head into it and took a long drink of milk. His head was in the container— a back leg extended—for about two minutes. He did not lap with his tongue; he seemed to be drinking as humans do, except that it was not from the edge of the container but from the surface of the milk in the center. He had two more nuts, one more swig of milk, and took off.

To make certain that what squirrels might consider a delicacy was not being denied them out of ignorance, I decided to offer the others milk. The next morning, a pair that appeared to be mates showed up. Each occupied a different corner of the fire escape to eat. She appeared to be uncomfortable in his presence and soon moved into the apartment to continue her meal on the bookcase and the sill. I offered both husband and wife the container of milk. Each sniffed it. She seemed a bit frightened by it. Neither drank any.

I wanted to learn whether Runty's milk drinking had been a whim or whether he was a regular tippler, so I got out of bed the next few mornings to greet the six A.M. cruncher. It was not Runty. He apparently did not come every day. When he did come one morning, I put out the container of milk. He moved off the sill when I set it there, but returned to eat a few peanuts and then take a long drink of milk. I must have neglected to fill the container high enough to suit him, for he tipped it, deliberately, to make drinking easier, spilling some on the windowsill in the process. I could neither see nor hear any lapping as he drained all the milk in the container. He made no attempt to clean up the sill.

A week later, Runty consumed a filbert in record time for him. I was about to cheer when he suddenly fled. Another squirrel had arrived. A female! Poor Runty, bottom man on the totem pole.

Runty must have been hiding somewhere in the wings; when the female left, he returned and fell to some almonds. He also fell from the windowsill, for the second time in several days. I observed, with pride, that he looked much better than when he had first arrived a month earlier. His face seemed less haggard, the bulge on his belly less pronounced; he was hiccupping less. He didn't drink any milk, and ate three filberts in succession in a respectable time.

May came. Runty came less frequently. I don't know why. Perhaps he had new sources of food: blossoms on the trees, people in the park to hand out peanuts. Almost three weeks went by without a visit. When he did appear, the first nut he took was a filbert. Eating time reasonable. He had several more, interspersed with peanuts. I put out his container of milk. He sniffed it but didn't drink. He had learned something in his absence: He backed out after taking a nut from the cup instead of falling off the sill in trying to turn around. He continued to back out each time he left the cup. Backing out became one of Runty's distinguishing characteristics.

We planned to be away in July and the beginning of August. This would be our first long separation from the squirrels. I had arranged for a squirrel sitter, Jeannie McCloskey, who had the apartment directly beneath ours. Jeannie assured me that feeding the squirrels would not be a chore and would have decided benefits. Jeannie had a large, old, neurotic gray Persian cat named Mouse, who was never allowed out of the house and had little to occupy his time. Mouse was fascinated by the squirrels. He went into Jeannie's bedroom every morning to watch the traffic on the fire escape. Jeannie said that some of the squirrels ate their nuts right outside her window, in full view of Mouse, effectively thumbing their noses at him through the protecting window or screen. I listened to Jeannie on this subject with benign amusement until I witnessed, in the garden, several instances of what must surely have been goading of a five-month-old cat, the tenant of the ground-floor apartment. First, there was walking in front of the cat, then a running from—not too fast—next, leaps into the air by both parties, with the final tableau showing the cat four feet up, on the trunk of a tree, and the squirrel on the ground, walking away.

I gave Jeannie fifteen pounds of assorted nuts, and ten dollars just in case, to cover the six weeks we would be gone. It remained to notify the squirrels of the change of address. I thought a good way was not to let them into the apartment a week before we were to leave, but to leave nuts outside on the windowsill, fewer each day. The last couple of days, Jeannie would put nuts out on her sill, so that there would be a short period when there would be nuts on both sills. Then Jeannie would take over.

The plan was working according to schedule, except for Runty. Runty was special, and needed special treatment. He came only a few times in June, but when he did, I gave up trying to cut down on the nut supply. Once, after he had been absent for three weeks, I noticed him on the fire-escape railing. By the time I got to the window to open the screen, his nose was pressed against it right where I was trying to slide it open. He marched in and took a filbert from my hand. He certainly had become bolder—and fatter. His tail still looked awful, though, and he still took more than average time to shell a filbert. I cracked one and put it, a peanut, and some milk near him. He ate the nuts but didn't touch the milk. I kept cracking filberts and he kept eating them. He had eaten nine nuts when another squirrel entered and Runty bolted. When the other one left, Runty came back to eat some more. I stopped cracking when I counted seventeen. I decided to let him crack his nuts himself, even if he was slow. The other squirrel came back while Runty was inside. Runty leaped over him in the entrance, despite being loaded down with seventeen nuts, and disappeared again—temporarily. (The entrance was now the screen, slid aside. The blocks of wood had gone with winter. A window filter, slid aside, had been the spring entrance.) Runty reappeared once more, when the coast was clear, to eat a few more nuts.

Twenty nuts singlemouthed. I didn't know whether this meant Runty hadn't eaten since I'd last seen him, whether he had a tapeworm, or what. At one point, when I was handing him a filbert, he got my thumb. This had happened often, with other squirrels, who were gently testing for the nuts they couldn't see directly in front of them. When a filbert, held between thumb and forefinger, is offered to a squirrel, he gently bites each finger until he comes to the hard, unyielding one—that would be the nut. The biting is necessary because these sharp-eyed creatures cannot see well close up. Their eyes are on the sides of their head—the better to protect them from predators, by providing a wide-angle view—and so, for objects close up and in front, they do not have binocular vision. The gentleness of the biting indicated to me that they are aware, in some sense, of the damage their teeth can inflict. But Runty, perhaps because of inexperience in taking nuts from a hand, hung on. Squirrels' teeth are sharper than kittens', but Runty's bite didn't puncture my skin because the sinking in was slow. I washed the spot thoroughly nevertheless, as I did not know what had made Runty a runt, and thought it might be transmittable.

We saw Runty again a few times after we returned in August. I worried when he didn't show up the first week. He came the second week though, and the third week, and a few times in September. We never saw him after that. We had not built him up into a two-pound terror, or a terror of any weight, for that matter; he remained frightened of others. Larry was troubled that we had interfered with nature in building him up to where he might be able to propagate his infirmities. I must confess, I was never able to summon up that amount of detachment. Runty was a pathetic little creature who needed my help; there were no other concerns. I thought about Runty in later years, when others came who needed help. By the time Runty disappeared, a couple of others had made their way into our hearts.

9. NOTCHKO & CO.

THE WOOD IN OUR WINDOW frame was crumbling. We were pretty sure that the damage was being done while we were away and someone wanted in. When we were at home and happened to notice a face peering in the window, we usually opened it, at any hour of the day, though technically, office hours were nine to ten. The fence of chicken wire I had rigged up to protect what was left of the window frame had prolonged its life a little, but every advantage appears to have its associated disadvantage. Loud twangs woke us early one morning as the chicken wire was assaulted. There was a tremendous lunge against the window in a drive aimed to go over the top. I asked Larry what he thought we should do. "Train them not to come so early," he mumbled, and rolled over back to sleep. I lay there worrying about what the sharp edges of the chicken wire might do. I had bent the top of the wire fence in toward the window expressly in order to prevent mishaps, but I had not counted on an over-the-wall assault.

When we got up and opened up shop, I taped the edges of the chicken wire and inspected all customers for injuries. There appeared not to be any. The chicken-wire fence went the way of the window box after a while. One character had taken to twanging it regularly each morning. As the twanging was not associated with an attempt to climb over, I interpreted it to be a deliberate noise-making operation, designed to attract attention and get breakfast moving.

The Eighth Street regulars accepted the plastic cup as theirs; they headed straight for it on entering, even when it was not in its customary spot on the windowsill but on the dresser or the floor. Only if the cup was empty would a squirrel wander around the room, looking elsewhere for a nut. However, sometimes we made them look elsewhere on purpose, so that we could have personal dealings with them. One Saturday afternoon, Larry and I were lounging on the bed, listening to WNYC's "French in the Air" on the pocket radio, which was lying flat on Larry's chest. Various squirrels trekked across the bed to take nuts from my hand. One walked across the radio, undisturbed by the French conversation beneath him. After a while, Larry rose and sat down in his chair. There, he was treated much like the radio. The squirrel took shortcuts across him to get to the bed, where I held the nuts.

I believe—though Larry does not—that the squirrels welcomed these personal contacts, and even sought them. Any number of times, a squirrel would take a nut from the cup and approach closer to me or to Larry to eat it. Sometimes the squirrel's face would be just inches from mine. One individual, called The Lady because her prominent teats proclaimed her a mother, liked me to talk to her. I would croon compliments like, "Such a *good* squirrel," at her, as she sat munching right near me. She lingered all the while I spoke to her. Occasionally, I would bring a stalk of celery into the bedroom so that we could lunch together. Once, I was certain she was either trying to answer me or trying to imitate me. She smacked her lips open and closed as she looked at me. I did the same. She did likewise. We kept on smacking our lips, The Lady and I. She was not chewing; there was nothing in her mouth. To this day, I don't know what our "conversation" was all about. I wonder what I might have said unwittingly.

I must confess that, of the first batch of squirrels, Chazzer, Hairy, and Slim are all a bit fuzzy in my mind. I have already admitted the difficulties surrounding Little White Ears. I was inexperienced when they came into my life. Even with The Lady, who was certainly a definite character all the time she was making her appearances—I took a picture of her lying on the fire-escape railing, holding in her mouth the nut she was too stuffed to eat; she looked like a suckling pig with an apple in its mouth—I couldn't tell exactly when those appearances started or ended. I did not yet know what to look for that would last. Runty was the one individual I could identify in court if I had to.

The second batch was the last one that was still fuzzy. That was the batch that taught me what—or rather, what not—to look for in a squirrel, what was only skin-deep. The second batch appeared one morning in May. It consisted, at first, of Yellow-foot and Calico. Yellowfoot, a male, had a yellowish tint all over, and a back left foot that was not yellowish but yellow. Calico, a female, had the mottled yellow coat of a calico cat. Why should I be fuzzy about such pronounced characteristics? Because fur is the one feature that changes most. In the spring, squirrels lose their thick winter coat. While they are losing it, some of them look terrible. The fur comes out in patches, even from their faces in some cases, altering their expressions. Some look like members of the Hari Krishna sect, when the only remnant of winter fur is a ridge sticking up along the spine. The winter fur appears soft and fluffs straight out from the skin. The spring coat is sleek and shining and lies flat against the body. In spring, squirrels suddenly look skinny, and you realize that the delightful round winter shape is a fake, all fluff.

Tail fur changes most of all. A tail can go from an appendage hardly more attractive than a rat's to a magnificent bush—or the reverse—in a few weeks. Some tails take on rings of color, others stripes from the center out, still others appear quite uniform; all show patterns different in a side view from those seen straight on. Tails sometimes even get shorter; this does not happen very often, and the change takes place not over weeks but overnight.

It is possible that Yellowfoot and Calico had been coming earlier unidentified under winter fur, and when spring came, showed their true colors, but I think not; they appeared to be juveniles. They overlapped with The Lady, who exhibited no yellow at all. About a week after Yellowfoot and Calico appeared, I met their associate, Notchko. I am much too serious-minded a person to believe in love at first sight. Attractions, yes; love, no— at least not with humans. For squirrels, I am not sure; it is not as if one must wait for an exchange of ideas to determine whether there is harmony.

I don't know what it was about Notchko that made me love him at first sight, certainly not the notch on his left ear—I didn't notice that until afterward. Probably it was his easygoing manner. His good nature was apparent at first sight, and not only to me. Larry and I both agree on that. There would be deep attachments with other squirrels, for different reasons—with Genius for his intelligence and eccentricity, for example—but with Notchko it was because of his temperament. Even as a youngster, he was mature and sophisticated, the kind of person you'd like to have as a friend.

Notchko, too, had a yellowish tinge, particularly on his face, which had a darker mustache-and-goatee type of marking superimposed, but he had no vivid yellow sections like Yellowfoot's yellow foot. He was shy at first, but rapidly became bolder, investigating the leaves of the rubber plant, lingering inside the apartment after taking his nut from the cup. He learned quickly: After knocking the cup off the sill, the next time he took a nut he steadied the cup with his paw.

Notchko liked to eat on Larry's shoulder. That started with his tapping Larry on the shoulder when the cup was empty and he wanted Larry, sitting in his armchair, engrossed in a book, to know. Larry would get up and go to the kitchen for filberts and Notchko would wait on the back of the chair for him to return. From the back of the chair it was an easy step to Larry's shoulder. His shoulder was also an easy place for Larry to get a whiff of Notchko's body odor—strong and acrid, quite unlike the warm, meaty odor of dogs—and for Notchko to look in Larry's ear. Larry would sometimes put a nut between his teeth for Notchko to take with his. That's how much trust there was between them.

Yellowfoot was Notchko's superior, even though they appeared to be the same age. Notchko was inside when Yellowfoot arrived one day. Yellowfoot got Notchko down on the floor and then, keeping himself between Notchko and the exit, wouldn't let him get out: Notchko rushed around inside, finally making his escape when Yellowfoot came after him and left the exit unguarded. I never saw Notchko do that to any other squirrel later on, when he became top man. With him it was always "live and let live, even ladies" (unlike Yellowfoot, who was not above attacking The Lady, who was old enough to be his mother—and may have been. It is true that her angry bleating, which started the moment she saw him, may have driven him to it, and that she did attack him first, but still . . .). And Notchko would never do what some of the other males did, namely, plant himself in the entranceway to eat, so that females and inferior males could not get to the cup.

As an adult squirrel, Notchko never budged from the status eating station at the edge of the fire escape, directly in front of the window, as he watched others come and go through the window. He didn't have to budge to see them; the bulging black beads that are squirrels' eyes permit an even wider field of view than is afforded simply by their position at the sides of the head.

(It amused me to know that a squirrel could be watching me even when he was facing away.) Notchko didn't even employ the ritual threatening lunge that some of the top males did. With him around, several individuals could eat at different spots on the fire escape—or even in the room—all at the same time, in uneasy peace. He was just a very nice guy.

Notchko's character was especially striking because, on the whole, the females seemed more sociable than the males. Females were the ones to take a nut and come closer to Larry or me to eat it, and it was a female who, in running from another squirrel, leaped onto Larry's lap for protection—"Daddy!" But those may have been statistical fluctuations. As I think back over the whole lot and pick out the truly disagreeable among them, there were three females and only two males.

10. *LIFE IS JUST A BOWL OF CHERRIES*

WE THOUGHT OF TRYING to get the squirrels to do tasks. No tricks —Larry was adamant about that. As a cat lover, he invoked terms like "independence" and "dignity." As a dog lover, I suspect that cat lovers invoke these terms to avoid admitting to themselves that cats don't have the intelligence to learn tricks. I am referring to small cats. The big cats, lions in particular, seem doglike to me.

We thought of having the squirrels get their filberts from a gumball machine, but we never pursued the idea. I did, every now and then, dangle the venetian-blind cord in front of a squirrel, hoping to initiate play. On each occasion, the animal would examine the plastic tassel at the end of the cord and, finding it not edible, would lose interest. Squirrels are quite curious about objects, but seem to have no sense of play—at least they had none with us. They would not retrieve. Anything thrown or rolled at them was investigated, eaten, or carried out if it was a nut, otherwise abandoned. One tried to pull my ring off my finger. He came across it on a routine inspection of my palm, got a good grip on it with his teeth, but as he kept tugging, not along the length of my finger but perpendicular to it, he did not get very far and gave up. Some of their chasing of one another must have been play, since there were no dire consequences, although it certainly did not look like play.

We finally abandoned all thought of tricks and tasks, partly out of laziness, partly out of the wish not to alter the squirrels in any way, but mostly because we found them doing a sufficient number of interesting things without our instruction.

Calico, for example, appeared to treat the entire bedroom as her own. While each of the others had a definite spot on the fire escape or windowsill that he or she habitually occupied while feeding, and that served us to identify which squirrel was present, Calico would sometimes eat on the desk, sometimes on the bookcase, sometimes on the bed, the latter usually when we were not in the room (we would find the telltale shells). Once I came into the room and saw Larry in his armchair, Calico near him on the desk lying flat on her belly like a starfish, her four legs and tail radiating out in all directions. She had apparently eaten so much that she needed a complete rest before resuming.

And Slim let me stroke him. While all the others to whom I made advances first squinched away from my hand, and then, if I persisted, socked me (Hairy went to the length of staying away for two weeks after I molested him, much to my anguish and subsequent relieved jealousy when I learned he was seeing Mrs. Russell) Slim once stood transfixed on the arm of my chair as I stroked him eight times by count, the expression on his face appearing to say: "What the heck is she doing to me? But whatever it is, it's not bad." Only one other squirrel let me stroke him repeatedly—Toughie, one of a pair of very young twins, the other being Timmie, for Timid. Toughie even shut his little eyes in ecstasy as I stroked.

I tried to touch all of the squirrels. I would sneak a finger out to get a feel when one came over for a nut. Some appeared not to notice.

Their table manners we found interesting. Squirrels are fastidious. While I (and most of my friends) eat the thin brown skin between a peanut and its shell, squirrels do not. Many do not eat the skin of a cherry, either. It occurred to me that they might be programmed to shell everything they eat, even when the "shell" is the skin of a cherry. However, that idea was ruled out by the observation that they do not shell a nutmeat that has already been shelled. And yet one of our squirrels (Philip, about whom there will be more later) would divest a shelled almond of its dark outer coating and eat only the white interior—a kind of squirrel analogue of glat kosher.

Whatever experiments we did perform centered mostly around food, which may reveal more about us than about the squirrels. Each squirrel would accept a single cherry, and peel it while eating it. There was no prior peeling, as with the shelling of a nut; there would simply be a mess of cherry peel and a pit left over after the cherry was gone. And no squirrel would accept a second cherry. Some, in fact, would not finish the first, but would leave half or one third, something which almost never happened with a nut.

The squirrels treated lychee nuts similarly. Lychees were not accepted out of hand. There was a prelude in which varying degrees of suspicion and reluctance were exhibited, depending upon the individual. Once they accepted one, they would roll it over and over before cracking it. They would eat the fruit and discard the pit. But each squirrel relayed to me the message: "Okay, I've eaten a lychee nut. I've had the experience. Now I don't have to have another one in my lifetime." And they didn't. Some didn't even complete the first experience.

I wondered how they knew which objects were nuts. Why did each squirrel take a Brazil nut without question? Certainly some, if not all, had never seen one before I offered it. And most could not hope to crack one without my assistance. Yet all grabbed Brazil nuts eagerly. I suppose it was the odor. Other foods produced reactions extending from disdain (grapes) to unconditional acceptance (walnuts). Apple fell in between, nearer to grapes; it was taken only in a pinch, and only by a desperate few.

There are decided preferences for certain nuts. After walnuts come filberts. I rarely furnished pecans, possibly because I don't especially like them, and because the squirrels seemed sufficiently spoiled without them. As Larry put it, they acted as if nuts grew on trees. Almonds were lower on the list, but higher than peanuts.

That is the gross outline of the nut hierarchy. A fine structure and variations are superimposed by special circumstances. For example, very young squirrels cannot crack filberts. They are delighted with peanuts. They also like sunflower seeds, which most adults have outgrown. And poor Halftail, the disagreeable female who ultimately won my sympathy, would eat virtually *anything* in the seed or nut class, her preference determined by how fast she could get it down. She liked cantaloupe seeds, as did several others, but she appeared to be the only one whose tastes stretched to the exotic: watermelon seeds.

The importance of nuts I measured by the willingness to carry them away. I suspect that importance depends upon both size of nut and hardness of shell. Hard-shelled nuts are exported for burial outside the apartment. Almonds are buried in the apartment. Peanuts are either consumed or ignored. Squirrels are programmed— or learn—that hard-shelled nuts last better in the ground than do soft-shelled nuts. I ultimately had to learn, as I am not programmed for it, that if I did not want to play the role of conduit between supermarket and ground, I should at least crack the nuts, as damaged nuts are not buried. I did not crack filberts, however, because I did not want to interfere totally with burying instincts. Besides, as there are more filberts in a pound than other nuts, each burial cost less and interfered less with certain instincts *I* have.

For economic reasons, I offered a macadamia nut only once.

It was seized in a way that would indicate the seizer knew the price.

Seed eating was fascinating to watch. I became an instinctive partner in fastidiousness. I would place the seeds on the window frame, which is an inch and a half or so above the outside window-sill. A seed eater would clutch a bunch of seeds to its mouth with front paws. All that would be visible to the outside was husks being fired out faster than one per second *onto the windowsill below,* the "table" remaining clean.

Enough talk of eating. There are higher things in life, even in a squirrel's life. And so we now turn to matters of the intellect.

11. GENIUS

I DIDN'T LIKE GENIUS at first. Like most geniuses, he had a disturbed personality. The name Genius must seem childish. I thought of calling him Isaac, after Newton. But then we would have to go through: "Isaac? He's the one that's so much smarter than the others, with the stumplike tail, etc." Besides, Isaac seemed a stupid name for a squirrel. And Genius didn't look like an Isaac. Furthermore, Larry didn't think he was *that* smart.

I am irritated by novelists who state that a character is brilliant or witty and then do not give that character a single brilliant or witty remark to say in support of the claim. Stendhal irritated me in that way in *The Charterhouse of Parma.* Since he was Stendhal, however, I blamed the absence of witty and brilliant dialogue on the translator. I now find myself somewhat in the position of one of these novelists in having named a squirrel Genius.

There are, let's face it, not too many ways in which superior intelligence can be exhibited by a squirrel. As one might expect, most involve food. Genius had many more burial sites than the others; he did not restrict himself to the pillows on the bed or the plants. His range was much wider within the bedroom. He used the narrow space between the filing cabinet and the bookshelves;

he would go to the rear of that space, creep under the bottom shelf or behind the books on the lowest shelf, and would leave his nut in a spot that was extremely difficult for me to get at—like the spaces between pipes in the radiator. Wide scope in sites might merely indicate adventurousness, but Genius checked his sites, and was the *only* one to do so. He would return to where he had deposited a nut the previous day to readjust and refine his earlier work, to be certain the nut was not visible. Sometimes he would move the nut to another site, particularly when the initial burial had been as impermanent as under papers. If the nut was not where he had left it—and I did my utmost to insure that it was not—his initial confusion would rapidly give way to a black look in my direction. In the interest of objectivity, I must admit that Larry was skeptical of the black look. However, I believe I developed greater insight into what went on in the minds of the squirrels than he did—I spent more time with them—and, as evidence, I offer the fact that Genius ceased burying in the bedroom. He abandoned the sites from which his stores disappeared and took to pioneering down the long hallway and into the living room to bury. He was sometimes as much as fifty feet from his point of entry.

At first, I was flattered by the trust I thought this implied. It was not trust, however. Nor was it the lack of common sense that ordinary people love to attribute to geniuses. It was boldness. Genius was the most intrepid of all the squirrels—with us, that is. He was a coward with his peers. He would force his way into the living room. (The word force is used advisedly.) Once, seeing that Genius would get through the bedroom door before I could close it, I shouted to Larry, in the kitchen, to head him off. Larry got to the hallway and placed himself between Genius and the living room. Genius veered to the right to get by. Larry sidestepped in front of him. With a swift bolt to the left, Genius shot by Larry. This would happen often, sometimes with three or four Chaplinesque dodges to the right and left before Genius would shoot by.

He got better all the time. We could not keep him out of the living room unless we shut the bedroom door. And why were we so anxious to keep him out? Because the smart alec was outwitting us. The hallway Dodg'em game was merely a prelude. The main feature went on in the living room, where Genius would spend as much as fifteen minutes sometimes, inspecting sites for possible burial. That, too, might perhaps be attributed to characteristics other than intelligence. Where Genius's intelligence beamed forth was in the burying itself. He made the usual burying motions, the gatherings in and the tampings down—*but he made them at three sites!* He had only one nut. He was trying to fake us out. We would spend hours trying to find where the nut had actually been deposited. Sometimes it wasn't recovered until Maude came to clean and it clanked into the vacuum cleaner. Several weren't found for years, not until everything was moved when the apartment was painted. Some, I fear, have not been found at all, and will one day sprout giant hazelnut trees. Genius buried under the sofa, in the sofa, behind the radiator, in the large copper vase with dried leaves, even in the stereo set, if the door was open. When the living room rug was replaced, during Genius's tenure, with one having a high pile, Genius had a field day. He buried right in the pile, not under anything—a possibility that had not entered into any of our purchase considerations. His task accomplished, he would head back to the bedroom to take another nut from the cup, or, if the cup was empty, to demand one of us. We didn't dare refuse him. Both Larry and I were somewhat afraid of him. His fearlessness intimidated us —and I believe he knew it.

With a new nut, the whole act would begin again, until *we* acquired enough sense to close the bedroom door before Genius got to it. Heading for the door, if he was still six feet from it and saw it being closed, he would turn on his heels and leave. Smart enough to know when to quit, he did not require the door to be shut in his face to know the game was up.

Genius was one tough squirrel. I learned that fact rather early, when I tried to pull him out of a plant in which he was burying. The fury with which he socked me left a lasting impression— visible for days.

As for his disturbed personality, that was evident in his general disagreeability. He was an unpleasant squirrel, almost surly. One could not hand him a nut. One had to throw or roll a nut to him. If it were handed to him, his grabbiness made it just as likely that a finger would be grabbed as the nut. No gentle checking which was the soft object—the finger—and which the nut. Genius had no time for such niceties. He simply grabbed. And if it was your finger he grabbed, too bad. I had warned my Rutgers colleague, Steve Bernow, that Genius had to be handled with care, when Steve came to meet the squirrels. There was a bit of a mixup though; I thought I was letting Steve hand a nut to another squirrel. I myself may have precipitated the unfortunate incident by shouting: "It's Genius!" startling Steve into pulling back. It was very embarrassing, Steve's being bitten on his first visit. As Steve was a brand-new father, with brand-new responsibilities, he had the wound attended to by a physician, who prescribed a tetanus shot.

I, too, was bitten and laid the blame to Genius. I am not referring to the minor bite on the cuticle I received during Genius's first week with us. This was a major bite in which a tooth went right through my thumbnail. It occurred during a lunge for a nut. I don't know quite what happened. I pulled back with an "Ow!," bleeding all over the filbert, and Genius stood still, clicking. When I had wiped off enough blood to use the phone, I called Dr. Ingerman, whom I had stuck with over the years, first because she never prescribed any medicines and second because I found out that her son was a well-known physicist—I figured the genes must be good and therefore she must be a good doctor. Dr. Ingerman thought I should get a booster tetanus shot—my last one had been administered thirteen months earlier—and recommended I go to a hospital emergency ward. The Health Department's advice was to catch the squirrel and have it examined for rabies, because the reason there had been no cases of rabies in New York in years was because all offending animals had been examined or the victims had received rabies shots; the incubation period was two weeks. Since Larry had been bitten a little over two weeks earlier and was not foaming at the mouth, I concluded I did not have rabies. Nevertheless, I called University Hospital; their recommendation was that I catch the squirrel, "Because nobody is going to give you the shots without a suspicion of rabies; they're prolonged, painful . . . and the incubation period can be as much as *two years.*" As for tetanus, see my local doctor was the advice; some think the shots wear off in one year, others in two.

I got the tetanus shot, but decided against the rabies vaccine, what with the disagreement between the Health Department and University Hospital, the pain of the shots, Genius being a member of the family—and how in heaven would I catch him? I wondered fleetingly if the taste of blood might produce a man-eating squirrel.

Now, I stated that the blame for the bite was laid to Genius. Actually, Genius was not, as a New York policeman would say, the perpetrator. The second diary entry about Genius records us contemplating the name Scary for him, for he seemed to rush around in terror, "sliding onto a nut on the sill the way a baseball player slides into base," and "unfortunately, he slid likewise onto Larry's finger, lunging at it to get the nut it held and producing our worst casualty to date. He's a peanut eater, possibly because he has no time to select and simply takes what is nearest. On further consideration, his name should be, not Scary but Misha, short for Meshuggeneh." The third entry started with "Misha is a genius," but the next entry states that Misha and Genius were two. They had both appeared around the same time. Misha had been the one to give me the bad bite. Misha hated me, I was sure. Returning in the evening, after having bitten me, Misha took one look at me and backed off, clicking. I saw hatred in the look. I saw that look of hatred each time Misha would enter and click.

The biter colored our relations with all the small squirrels, because we couldn't be sure who was who. I was overjoyed to find out that Misha and Genius were not the same guy. In fact, Misha turned out to be a girl. Misha was the one that did the serious biting and sliding into base—probably out of terror. Genius grabbed, but did no major damage. I forgave him his unpleasantness. I tend, much to Larry's consternation, to forgive the trespasses of human geniuses as well.

I should add that no squirrel ever bit us out of malice. Bites were almost always accidental. I can think of only two instances when the bites were deliberate, and in both instances there was provocation. In one, I was trying to "train" the offender—a story that will be told in detail later—and in the other, Larry held his hand over the cup of nuts while I was getting set to take a photograph.

Misha, Genius, Runty (after buildup), and possibly one other squirrel, all looked alike. They were small, primarily peanut eaters, and did not have bushy tails. I studied them more carefully. Misha and Genius had tails that were round in cross-section. Runty's and the other's were flatter. Misha and Genius had markings that looked like mustaches, but Genius's face was slightly yellowish. Runty had his hiccup, of course, and a peculiar way of holding his head up high. Genius's tail was truncated. It had the appearance of a club. It looked as though it had been chopped off. It was almost as long as anyone else's, but it was squared off at the end; you could see tail, rather than fur, if you looked at it straight on. If it was not immediately apparent with whom I was dealing, I learned to look for gender—the foolproof method of distinguishing Misha and Genius.

In most instances, subtlety was not required in identifying Genius. Larry and I might be in the kitchen, having lunch, when we'd see a squirrel marching past the door: Genius on his way to the living room. Genius was tough and bold, not afraid of anything in our house, not even the dark (which squirrels appear to avoid). He crept under the bed to deposit nuts. None of the others ventured there. Perhaps his eyesight was better. Once, when I wasn't expecting him, I almost stepped on him in the semidark foyer next to the living room. I recoiled. There was no thought process; I just recoiled from a large, live object. A rat? When I realized it was Genius in the gloom, I felt like apologizing to him.

If he was not very talented with his hands to begin with, Genius became quite adroit, unlike the stereotype of the physically awkward mental giant. I may even have unintentionally contributed to his dexterity by rolling filberts at him. A clumsy catcher at the beginning, he got better and better, until he could keep a filbert from rolling off the windowsill with a paw, and one day caught a filbert on a bounce. He would scoop nuts out of the cup with a paw; the others used their jaws.

One of Genius's idiosyncrasies was his dislike of physical contact (which is another reason I wanted to name him after Newton). When I worked up enough courage to offer him a nut from my hand, he would take it, but only with hesitation. He would not come onto the arm of my chair, let alone onto my lap. He was also intensely private; he did not like to be observed while he ate. If he happened to be dining on the fire escape and I came to the window to say hello, he would move out of my sight onto my neighbor Yoko's windowsill. This latter trait he shared with several other squirrels, however. (The more sociable ones, Ninotchka for example, would do just the opposite; her response to a hello would be to have her next nut inside, close to my chair, particularly if I continued to converse with her.)

Genius was quick to learn what a loud "No!" meant—usually emitted when a nut was about to be buried in a plant. Once, his not obeying a "No!" and my not respecting his dislike of physical contact—I grasped him around the waist to pull him out of a plant —resulted in three parallel scratches on my hand from his nails. But I had been correct in my assumption that I would be safe from a bite because his mouth was occupied with the nut he was about to bury: Nuts are never dropped deliberately except in dire emergencies.

Genius was very much aware of his surroundings. He stood up tall to look into the eyes of a bust we keep on a bookshelf in the living room. Any change in the apartment, within his territory, would be checked out. Yukap Hahn, Larry's Korean postdoc, gave Larry a large red enamel vase, inlaid with mother-of-pearl, for his birthday. I placed it on the floor in the hallway, just outside the bedroom door. Genius passed it on his way to the living room, stopped, looked it over, and continued on his way. On his way back, he gave the vase another once-over. One more inspection and it was an accepted part of the scenery. When I moved the rubber plant from the right to the left side of the window, to give neglected leaves their place in the sun, Genius would be quick to note the change. If he knocked a nut off the sill, he would look where it had fallen. If he knocked the cup off the sill, he would watch it until it stopped rolling, sometimes jumping down to the floor for closer observation. I know that others had done similar things; it was the *sum* of Genius's enterprises that set him apart. In fact, at the risk of appearing sexist, I should note that several of the males, when new to the premises, detoured from the road to the cup to spend some time exploring. Some just looked—at the venetian blind, the floor, the rubber plant—others touched—the blind cord, the plant leaves—one chomped on a couple of leaves, punching permanent holes. They were always males, though, in my recollection. Perhaps the male squirrel is the explorer.

One of Genius's interests froze me with fear. When I opened up shop in the morning, sometimes it was straight from bed and

I was barefoot. Genius was fascinated by my toes. I never did find out whether it was their resemblance to filberts or their odor or whether he was a foot fetishist, but he would put his cold nose on one toe after the other as I stood motionless, waiting for the inspection to be over, fearing movement might produce mayhem. As a corollary to his interest in toes, or an extension to a general interest in extremities, Genius gave Larry's fingers an investigatory nip once, when he sat reading with one arm dangling.

At first, I didn't like Genius, but he interested me. He was a genuine individual. It was upsetting, though, to see him picking on females in a particularly offensive way. He made the threatening gestures all males made to females, but Genius followed through. He actually attacked, sinking his teeth or nails into their backs. It was hard not to compare his method of ruling with Notchko's. Genius ruled through fear, where with Notchko it had been respect.

Genius had altercations with me as well as with squirrels, apart from the bitings that occurred at the beginning of our acquaintance, which had been, by and large, accidental. Once, when I gave him a bad nut, he clicked at me, as if I had done it on purpose. Another time, it was a fight over acorns. That occurred during one of those difficult times in September when supermarket stocks of nuts are unreliable. Genius had been eating acorns imported from Westchester, with relish, I thought, especially those with sprouts on top; he would chomp off the sprout first. I mentioned the difficulty in getting filberts in front of a Lebanese student, who volunteered to bring in some he had; he didn't like them uncooked, he said, and was too lazy to cook them. They turned out to be pecans—I never did learn how he cooked them—and rather old ones at that. Larry didn't think we should give old nuts to the squirrels, so he ate them himself. The result was: I had only acorns for Genius and he wanted filberts. I put acorns on the outside sill and shut the screen. Genius was furious. He lashed his tail and clicked for several seconds, then tore off the plastic tape I had put over the chewed window frame.

Dealings with Genius were like dealings with a colleague, someone who is not exactly a friend but whose habits and personality are familiar. Genius and I each had rules of behavior, which the other observed. Genius knew what the sound of the alarm clock meant. Sometimes I'd be up before the alarm went off and could see him peering intently in the direction of the bed, his paw on the screen, near his face, giving the impression he had his hand over his eye to cut out glare and see better. Or I would hear him pacing back and forth. One New Year's Day, we didn't awaken until ten-thirty. Genius hung around outside from nine until then. When the alarm went off, there was wild excitement. He was free to make all the noise he could, which he did mostly by walking on the window filter. January 1 is Larry's birthday, and it was walnuts all around, so the wait was worth it for Genius.

Ordinarily, if he arrived after nine-thirty and found the entrance closed, he would climb up the filter (the screen in summer) and tap on the glass of the window with his nails. I believe he did it to attract my attention, because that was its effect. I can't say whether he did the same when I was not there.

Always impatient, he took to helping me open the filter or screen. He would push with his paws on the spot where my fingers were pushing. Occasionally, he would bite the window frame as well. I had to open the filter and screen rather carefully—I had learned, painfully, that the eager anticipation of breakfast resulted in anything extended out the window being treated like a nut.

Once, when I wanted to leave, I gave Genius a filbert and started to close the screen behind him. He stopped, nut in mouth, and watched my every movement. I could tell he was studying how it was done. I could tell this because, a few years earlier, Scarface had stared dumbly as I closed the screen. I was under the impression that he was somewhat dense, but then, after a few days of staring, Scarface placed his back paws on one side of the screen, his nails hooking the wires, put his front paws on the other side, and yanked, sliding one side across the other. Scarface opened the screen on his first try, after a theoretical analysis. Genius finished his filbert and did the same. In order to foil both of them, I poked a paper clip through the overlap section of the screen, so that one side couldn't slide across the other.

Genius came regularly. There was a period of time, though, when he would take only one or two nuts per day. It occurred to me that he might be doing exactly what I was doing. That is, I was putting out nuts every day, even when there was plenty of food available in the park or our garden, in order that the squirrels not lose the habit of coming round. Genius might be thinking—in whatever terms he used—"I must put in an appearance here every day, even when I don't need food, or she'll get out of the habit of putting out nuts."

Genius trusted me. One winter morning when it was snowing, in addition to using his tail as an umbrella (the word squirrel is derived from the Greek *skiouros*, "shadow tail"), Genius took shelter and ate right in the entrance, directly under the window. To keep out the cold, I lowered the window as much as I could, slowly, of course, until it just touched Genius's head. He did not stop chewing for a moment. The incident is also an indication of squirrels' concern for the weather. Genius subordinated his strong need for privacy while dining to his dislike of snow.

I liked the way he—and all the squirrels—shook the moisture off their fur, not the way a dog does, rotating back and forth about a horizontal axis nose to tail, but more in belly-dancer fashion, by a spiral motion about a tilted axis from the rear in near-sitting position to front paws aloft, like a screw being unscrewed.

Despite Genius's trust in me, I never did get to know much about his love life. I do know that in January 1974, he was in love with Ninotchka.

It was easy to be in love with Ninotchka. Larry was in love with her, too. She was easily the most lovable of all the squirrels. She never merely came into the apartment, she would burst in. Incredibly graceful, she would leap from the sill onto the back of the desk chair, pause, trembling with vitality, eyes wide, her entire body pulsating, proclaiming eagerly, "I'm here!" "Bright-eyed and bushy-tailed" is the perfect description for Ninotchka. She was more full of life than any creature I'd ever seen. Ninotchka got her name from the small notch on her left ear, smaller than Notchko's. The Russian name suited her; her *joie de vivre* made me think of Natasha in *War and Peace.* She was with us for several years, and probably was the sister of Round-the-Grill (named for the path she invariably took to and from the cup of nuts). The two arrived together as youngsters. Ninotchka was a mischief-maker from the beginning. As a child, she had stolen the cup of nuts. I saw her backing out the entrance with it. She took it partway up the fire-escape stairs to the roof, about ten feet from our window. Recovery proved a mess, because it involved going about six feet along the narrow section of the fire escape bordering the open, unrailed staircase area. My "edge complex" got the best of me. Larry had to get the cup. We tied it to the burglar grill after that. Ninotchka was also responsible for a major chewing operation on the bottom of the window, an effort to break in that we "repaired" —and prevented recurrence of—by using a long iron bar as a barricade. Ninotchka was such a charmer that we forgave her everything.

Ninotchka and Larry developed a special relationship. She liked to sit on his lap while she ate. By actual timing, once she sat there for forty-five minutes while she consumed about twenty-five filberts. She was pregnant at the time.

I like to think I am sensitive to the subtle signs a human being gives of being in love with another, before there is public acknowledgment. With squirrels, I am not so sensitive, but the signs are less subtle.

On a cold day, with snow all over, Ninotchka arrived late and began to bury filberts one foot apart around the edge of the fire escape. It was funny to watch her struggling to cover a nut with the powdery snow; it wouldn't stay covered. She had to gather quite a bit of snow from an area larger than when she covered with earth or with her imagination.

She gave me a fright when she slipped on a patch of ice on top of the railing; but she caught herself, this Olga Korbutt of squirreldom, and bounded into the house and onto the arm of my chair and then onto my lap. Genius was right behind her. She must have known he was there, but gave no sign; she calmly accepted a nut from my hand and went out with it, Genius inches behind her. Genius had not taken a nut. Since females normally run from males at mealtime, Ninotchka's calm told me she was aware of the power of love.

Genius followed his ladyfriend around for days. I never found out whether Ninotchka requited his love, whether it was an actual affair, or how long Genius remained smitten. There was a rival, a young fellow called Holey (for the little hole in his right ear) and there also seemed to be another female Genius was interested in. All very messy. A veritable Bob and Carol, Ted and Alice.

Toward August, I thought Genius had a harem. He arrived with two females for several days running. By November, Genius was entirely over his passion for Ninotchka and expressing interest in Push-off. (In turning back from the cup, after having taken a nut, she would push off the grill the way a swimmer pushes off the end of the pool to start a new lap.) They were performing that slow chase I had come to recognize as courtship: along the fire-escape railing, down a vertical bar to the floor of the fire escape, across our outside sill, and up onto the railing again, she in front, he right behind, his nose an inch from her rear. Then she came inside and climbed up the grill. He looked up, but didn't follow. He went to the cup. One appetite at a time, I guess. They took turns at the cup. When it was empty, Larry went to the kitchen to refill it, accompanied some distance behind by Genius. When both returned to the bedroom, Genius still on the floor, Push-off poked her head in the entrance. Genius made a rush for her. She retreated, and he departed. She came in quickly, hurriedly snatched a peanut, and went after him, reinforcing my impression that they were lovers.

I got the impression that relationships are pretty fluid. A twosome one day would be a *ménage à trois* the next. Harem today, gone tomorrow. I became involved, disapproving, parentally, of the pursuit by an older male of a female I thought too young.

Genius had grown from a slim adolescent to a solid, sturdy fellow. He looked like a tank, determinedly trundling across the windowsill to the cup. He was recognizable from across the room. He had had a few of what I suppose are the analogue of standard childhood diseases. He appeared to be susceptible to colds; there were several periods when he sniffed a lot. I could even recognize him by his sniff at those times. And once he had a urinary problem: He made odd, horseback-riding sorts of motions, and when he came to a halt there was a spot of liquid on the sill. If what he had been doing was urinating, it was in a squatting position. (The only urinating we'd ever seen by any individual was from a squat. I didn't think to note at the time whether it was a male or a female.) Anyway, Ninotchka was interested enough in the spot of liquid to sniff it.

One morning in the beginning of January, a year after the affair with Ninotchka, we heard a rattling of the window filter before nine. When we opened at nine, a squirrel that looked like Genius and had his head down, slouchy walk, entered. But this squirrel wheezed and sniffed badly, almost continuously, as if he were unable to breathe and were trying to force something out of his nostrils. It was the worst cold I'd ever seen—or heard—in a squirrel, if a cold was what it was. The squirrel took a nut from the cup and went to the leftmost position on the fire-escape railing to eat, a spot not normally occupied by Genius. I concluded that it was not Genius when I noticed that the squirrel's right ear was split from top to base.

He was in again at nine the next day, sniffing almost as loudly as the first time but not wheezing at least, and occupying either the leftmost spot on the fire-escape railing or a spot on the outside sill, just opposite the cup. I could see him clearly in the second location. It *was* Genius. His ear was in bad shape. The vertical split was not a clean cut, it was a tear. His nose was running, he sneezed a great deal, and his right eye kept filling with fluid and shutting. He repeatedly licked his left paw. I hoped he might be grooming, but as I had never seen Genius groom himself before, I knew it was more likely that he was licking a wound. I suspected, from the fear he showed of Holey, that it was Holey who had put him in this sorry state. Despite his eye, ear, nose, and throat problems, Genius's walk and manner were the same as ever. He pushed right past my proffered almond to take an almond from the cup. His appetite was unimpaired.

As the days went by, his nasal condition improved. The damage to his psyche, however, was slower to heal. When another squirrel came into view, Genius would bolt, even though several times the other turned out to be a youngster or a girl. Round-the-Grill slipped in while Genius was occupying his new spot on the outside sill opposite the cup, and Genius made no move to stop her. He was evidently not looking for fights. On another occasion, I heard Genius barking, then making high-pitched, gibbering sounds, and he entered hesitant and frightened.

One morning, Holey had been in early and, because it was raining, had eaten on the sill inside; ordinarily, he preferred to breakfast out of doors. When Genius came in later, he sniffed the spot on which Holey had sat, and then sat down there himself. What this meant, I do not know. My presence nearby soon annoyed him, and he left for his outside spot. When the cup was empty, I handed him a nut. I thought he seemed gentler than usual, but still rougher than any of the others. He touched my hand with his paw as he took the nut with his teeth. It was a rough touch, more like a push. I got a good look at his ear. It was now in three parts, like a script letter "m."

By one month after his ear-splitting experience, he was back to biting Charcoal, a dark little female. I saw him give her what I thought was a bad bite, but as she did not appear to be suffering, I concluded that the bite may have been some male-female ritual. The conclusion was reinforced when they began appearing in the house together, and when Genius bit her a second time without result.

On the whole, however, I had the impression that Genius was in a period of decline. He had been intimidated by males before; I had attributed that to his being an adolescent. The only males he ever chased were maimed in some fashion. But he was now a big, hefty fellow, and still intimidated. And it was not only Holey he feared. One afternoon in late February, Genius climbed up on the window filter and made a racket. As he did not normally do that if he had had his morning rations, I surmised that he had not had anything to eat that morning, possibly to avoid coming around when others would be there. I opened up for him. But then Patchy appeared. Now Patchy looked as bad as it is possible for a squirrel to look. Large patches of fur were missing, and as he was black, the bald sections showed pink. To be mottled pink and black might indicate no more than early moulting and that black squirrels moult unattractively. But Patchy had a tumorlike growth on his furless tail. Fearing he might have a skin disease, I had not sought to overcome his natural shyness; I never handed him nuts but placed them outside for him. He had yet to come into the house. I felt sorry for Patchy and put down a peanut for him, which he refused. I put a filbert in front of him; he took that and proceeded to shell it. Meanwhile, Genius, who had been having high tea on his railing spot, was ready for another nut. But Patchy, who up until that day had been maintaining a live-and-let-live attitude toward Genius, suddenly stopped Genius from coming in. I resolved not to encourage Patchy, for fear of losing Genius.

It was shortly after this encounter with Patchy that Genius gave Charcoal the second bite I witnessed; it occurred a day after Charcoal appeared outside with Patchy. I make no assumptions about a possible connection between her appearance with Patchy and the bite. It was also about this time that I became aware of how much Genius meant to me. This surly, now sullen creature had become very dear to me over the almost four years that he had been calling. He was becoming more of a recluse every day, and more of a coward as well, disappearing in the middle of a meal when someone else appeared, and returning only after the other one left. Now, even Ninotchka, our darling, who admittedly terrorized all the females, could frighten Genius.

He still did things that made me chuckle, though. He continued to lunge at whatever was in my hand. In one of these lunges, he got a peanut in his mouth before he realized it was a peanut at which he had lunged; he didn't want a peanut. Another time, after we had gone through the routine of his heading for the living room, my shutting the bedroom door to prevent it, and his having turned back six feet from the door, he returned a little later to find the door ajar and me some distance from it; he *sprinted* through before I could get to close it. And once, I *saw* Genius make a decision. He had carried outside the last filbert in the cup. When he returned, he took a peanut that lay next to the cup and headed out. A few steps down the sill, he hesitated, turned back to the cup—as if thinking: "Maybe there's something better in there"—took two steps, turned once again to the exit—"Naw, it was empty a minute ago"—and out.

I decided that I had been unfair to Genius, and that I should modify my statements about his grabbiness. He would clutch my hand with his paw when he took a nut from my fingers with his teeth. I did not feel his teeth; I felt his claws. It occurred to me that this might be the result of Genius's using his hands more than others did. Since he scooped nuts out of the cup with a paw, rather than with his mouth, might he not simply be using his paws at every opportunity, including clutching the hand that fed him?

12. TEEN-AGE HOODLUMS

IN AUGUST WE RETURNED from the West after an absence of two months. I unbarricaded the fire-escape window the first morning we were home, and was disappointed that there was no mad rush to greet us. The barricades consisted of a closed and locked window, at the bottom of which was the long, right-angled bar that had been installed to counter Ninotchka's onslaught, the filter immediately outside the window to provide an extra obstacle if the screen, which was right outside the filter, was eaten through or removed despite its having been made unslidable by paper clips and its being tied with picture wire to iron bars, on the side of the window, that had at one time supported awnings. The barricades were not the product of neurosis; the first summer we went away, there had been such hysterical rattling of the screen, before all the squirrels discovered the food supply at Jeannie's, that Yoko had to check that it wasn't humans forcing their way in.

When I saw Jeannie, later in the morning, she reported that all was well. Genius was fascinated by Jeannie's cat, Mouse, and would enter her apartment—he squeezed through the inch of space between the screen and the window, which he could do because Jeannie didn't keep her screen directly under the window, but had it in the outer track—and would climb up her burglar grill to the curtain rod, from where he would study Mouse at length. He did this several times. Either he thought he had time to escape if Mouse should go after him, or that old Mouse was not much of a threat.

Jeannie also reported that the squirrels had been fed that morning—roasted soybeans. They loved them, she claimed. (I had never succeeded in getting a single individual to sample a soybean.) The reason for the soybeans was not that the supermarkets were out of nuts—the squirrels had gone through twenty-six pounds in the eight weeks we were gone—the reason was Superintendent Lassiter. He had forbidden Jeannie to feed the squirrels, alleging they had chewed off parts of the windows of several apartments, one female having gone so far as to chew her way through an air-conditioner partition to get into Camille Strom's apartment, where she delivered herself of a litter on the bed. Jeannie had been disobeying. Soybeans left no shelltale evidence.

I was skeptical about the babies on the bed. Mr. Lassiter has been known to embroider a story for effect. How could the female —probably Ninotchka—pick precisely the apartment that was temporarily tenantless? As Mr. Lassiter was on vacation, I decided to go on with regular feeding while I worked out what to do.

In addition to the annoyance of the feeding interdiction, I had the feeling that the squirrels were unfamiliar. At first, I attributed this to my long absence from them. But Genius's ear and Ninotchka's exuberance were unmistakable; they simply were not calling. Instead, there seemed to be at least three very young squirrels. One had a tail half as long as normal, which she kept grabbing in order to groom. She was the neatest—or dirtiest—squirrel we'd ever had. For the first few weeks, she would groom herself every time we saw her. She would paw her face the way a cat does. Then she would bring her tail forward between her hind legs, run her front paws along its length, and lick or bite or just wet the end. She also licked and rubbed the rest of her body, but not as much as her tail. There were also two males, one gray with a long, narrow tail and one nearly black. As they were all the same size, I thought they might all be from the same litter. My summer reading had taught me that the blacks are a melanic variant of the grays. I suspected that the father of the litter was Genius. Although the long-tailed gray—henceforth Longtail—in no way resembled the suspected father, Halftail's half-tail had the stumpy, chopped-off appearance of Genius's and the same tuft on the end. And Halftail had Genius's temperament. She was extremely grabby and would leap onto a nut with a growl. She also leaped onto my hand with a growl, and grabbed my thumb—when I was attempting to assist her by tossing her an almond because the blackish male wouldn't let her in. The blackish male was even nastier than Halftail. I saw him eating a nut on the extreme left position on the fire-escape railing when Halftail appeared on the outer windowsill. He jumped from the railing onto Halftail and bit her—his sister. He even had the gall to challenge his putative father, Genius, who finally showed up after I had fretted for almost a week. Genius was inside, about to leave, when the little blackish one entered, jumped over Genius, landed in front of the cup, and grabbed a nut.

It was all very unpleasant, supporting squirrels we didn't like. Worse, they kept away those we did. Besides Genius, only Round-the-Grill and Curly Ears had been in. I decided not to feed this gang of youngsters. If I were to discourage them, they would learn to fend for themselves. Besides, they were arriving as early as six-thirty in the morning.

My plan was to change the normal operating procedure by not keeping the screen open; instead, I would open it only for those we knew and liked. Curly Ears arrived first, the morning I put the plan into action, and appeared to be alone. I slid the screen open for her. She had eaten several nuts when one of the marauders arrived. I slid the screen closed. It was Longtail. He scratched the screen a bit, but soon went away. When Round-the-Grill arrived, I opened for her. I heard growling. The black-ish one, a mere child, was chasing Round-the-Grill, an adult! I closed the screen. He returned from his chase and wanted in. He climbed all over the screen and up to the glass of the window. He did not use his teeth, though, and after a while, left. Round-the-Grill returned, found the screen shut and, seeing me inside, pawed the screen delicately. When that brought no immediate response, she began to bite the vertical wooden slat on the win-dow frame, against which the screen rests. I shouted for her to stop, and hit the screen when she didn't. I also, stupidly, opened the screen and gave her a nut, realizing as I did so that she would assume that biting gets the screen open. No sooner did she exit when the little blackish one, who had evidently been lying in wait, rushed in. The cup was not in its usual spot, so he found nothing. I made to shoo him out. He made a threatening gesture and growled at me! This was the toughest squirrel of the lot; even at his tender age he was making Genius seem like a sissy. I yelled, "Get out!" He ran around the room, wildly jump-ing on everything, including Larry. "And he's only a baby," I said. "Can you imagine what he'll be like when he grows up?" "Yeah," Larry answered, "a rhinoceros." Finally, he ran out and I started to close the screen. He turned and tried to squeeze back in. Fortunately, another squirrel appeared on the fire escape just then; he had something to chase.

That was not the end of the episode. He returned a few minutes later, while I was taping the chewed slat, and again tried to barge in. I shut the screen in time. When he left, I sprayed the entire area with Puppychew, a chemical advertised as harmless but having an odor offensive to puppies. I hoped squirrels would feel the same way about it.

Matters were getting more and more complicated. Larry and I were heading for the kitchen for lunch when Genius appeared at the window. I let him in. The cup was empty, however, and so I continued on into the kitchen to fill it. Genius followed me all the way. I handed him a filbert. One of the good things that was coming out of all the mess was that Genius was responding to the special personal treatment he was getting with a gentleness that was new to him.

I shut the screen and waited in the bedroom for Genius to finish his filbert on Yoko's windowsill. I judged by the sound; Genius's dislike of being observed while dining had intensified to the point where he now regularly ate out of sight, but within hearing distance. When he returned, I slid the screen open and handed him another filbert, which he again accepted gently. Larry was making lunch noises by then, and so I put five filberts outside and closed the screen, in time to notice Genius take one of them away with what appeared to be distaste—the nuts had lain on a spot I had sprayed with Puppy Chew—and the little blackish fellow trying to transport two of the remaining four. I had thought the little hooligan might return, but I had also thought Genius was big enough to defend his cache. Genius had either left, or was practicing discretion—or cowardice. The little one dropped one of his two filberts, as was to be expected; I retrieved it and the remaining two and went in to lunch.

It was evident that there would be no simple solution to the problem. After lunch—mine—Genius was back. I opened the screen for him, handed him a nut, and shut the screen after him. He was puzzled. He stared at the shut screen. Two of the little ones appeared, Blackish and Longtail; they went away peacefully. Genius kept returning. I drew my chair closer to the window so that I could interrupt my reading with less effort. Genius seemed less bright than his name warranted. Not only did he show little understanding of the situation as a whole, but he exhibited some specific denseness as well. When I let him in, I would place one hand on the center of the screen and slide the right-hand section away from the wall and across the rest of the screen with my other hand, creating an entrance on the right. Each time, Genius would go to the center, where my hand was, even though he would ultimately enter at a point two feet away. Perhaps his confusion resulted from his having to check out all the new aspects of the operation—or was I simply making excuses for him? I had to hurry each time I saw his face at the window; he scratched the screen when I was slow. The screen was already suffering from excess assistance in opening it. Genius had dug a hole in the mesh large enough for him to fit through; I had taped it and moved it to the corner diagonally opposite the entrance by turning the screen upside down and inside out.

It was all very upsetting. Genius was confused, and heretofore well-treated, well-fed young squirrels had, in one blow, learned what life is like. They resisted the knowledge. They climbed over the screen, making a racket early in the morning. Then I had to synchronize brushing my teeth with Curly Ears's breakfast habits. Fearing a little one might arrive while she was eating, I shut the screen after her each time I gave her a nut, just as I did with Genius. Unfortunately, Curly Ears was a slow eater, and ate more like a horse than a squirrel. It took the better part of half an hour to attend to her, if I was to be at the window when she was ready for another nut. All told, I was losing about an hour each day hanging around the window while Genius, Curly Ears, and Round-the-Grill ate leisurely breakfasts. And I was losing sleep out of guilt.

The situation continued unchanged for several days. Genius and Curly Ears appeared to accept the new regime; they gave up scratching the screen when it was closed. In fact, I was impressed with Genius's obedience; he did not attempt to break in. I wasn't worried about Round-the-Grill; in the previous few months, she had been burying madly; she could hardly be going hungry. The little ones depressed me, though. They kept hanging around, checking, checking. Blackish repeatedly tried to break in. And I was still being awakened early by some individual who couldn't believe this bounty was no longer forthcoming. I had an awful nightmare one night—extremely rare for me—in which squirrels were crawling over the screen like giant roaches.

Everyone was irritable. The irritability erupted in a fight between Genius and Larry. Genius had been in that morning, had eaten ten nuts, three more than his customary seven—though a modest number compared to Curly Ears's fifteen—so when he showed up toward evening to get nuts for burying, we weren't very sympathetic. He began to bite the slat of the window frame against which the screen rested. When he paid no attention to a shouted command to desist, Larry hit the screen to drive him away. Genius hit the screen, too. Larry hit it again. Genius hit it back, clicking furiously. Larry outlasted him. Genius went away, returned a few times, then left for the night. I worried that his anger might be permanent.

On balance, the system had been successful. By the end of September, the little blackish marauder was no longer coming, the other two were coming much less frequently, and, the blow-up with Larry apart, Genius had become positively genial. There were a few more episodes involving the two marauders before they became civilized. Once, when Longtail came in to steal a filbert, I tried to close the screen when he'd gone out. He saw what I was up to and tried to squeeze back in while I manipulated the screen. It caught him at the waist and I couldn't close it farther. Longtail finally backed out.

13. GENIUS
IN DECLINE

When Genius had been with us for four and one-half years, I began to notice distinct changes in him. I am not referring to the loss of the back part of his ear, though Genius's ears had certainly suffered major changes. After starting out perfect—if slightly pointed—one had been divided in three parts. One of those parts having been lost somehow, a two-part sawtooth remained for a time, after which the smaller, rear section of the sawtooth had disappeared. (Other squirrels' ears must have undergone similar metamorphoses, so my method of identification by ears was suspect. Halftail arrived one morning with a swollen face and a split left ear. The swelling went away in a few days; the split remained. What about the others? Was Ninotchka, who had been missing for some time, mascarading under a new ear? Might she not be Curly Ears, who had a tiny notch in the same place as Ninotchka's? They were alike in jumping onto the desk. But Ninotchka's ears weren't curly. Ears couldn't one day curl, as if from a permanent. Could something be frightening enough to make a squirrel's ears curl?)

Genius was undergoing deep personality changes. He began to act middle-aged. He started to do some of the things I had seen older males do: He looked around before entering the apartment; he'd stare at us and at the surroundings. One day, he tried to carry out two filberts, something a mature squirrel like Genius should have known better than to attempt. When I offered him an almond, he was extremely gentle in accepting it, touching me rather than shoving or socking me. He began to eat peanuts, his preference in adolescence. He developed a new idiosyncrasy, a next-to-last peanut routine in which, when only two peanuts remained in the cup, he would place one next to the cup, carry the last one out of the apartment, and return for the one he'd placed beside the cup. After a while, he would eat *only* peanuts. As I had recently changed *my* routine and had begun placing peanuts on the windowsill and filberts in the cup, I started to suspect that Genius had been forbidden to take nuts from the cup. That thought entered my mind when I saw him on the desk at the ashtray that held nuts for Curly Ears, for whom I *knew* the cup was off-limits. Genius would occasionally go to the cup—when I was blocking his path to other sources, for example. Even though he had mellowed, he had never become friendly. He persistently ignored me. I would hold out a filbert to him and he would jump past my hand to get a filbert from the ashtray on the desk. Sometimes, when my hand was between him and the ashtray, he would take a swift peek into the cup—nothing there—and then deign to accept the one in my hand. Even when he was compelled by rain to eat in the house, he did his best to ignore me, to respond to none of my calls or enticements. And somehow, I began to love this independence. I didn't want to think it might be dislike of me. Even when he tried to avoid being seen, the most I would call him was unsociable.

Genius was a complex character, a true individual. Where once I thought him surly, I now found him moody, interesting. On the few occasions when he took nuts from my hand, it was still with a lunge, but he had learned not to scratch. In his impatience, he would lunge while I was still approaching him, so that although his back feet were still on the sill, his weight would be almost entirely on my hand, a foot away from the sill. He was heavy. He would have fallen flat on his face if I had not been supporting him.

In late spring, I began to worry about Genius's ego. When a big, skinny male appeared on the premises, I worried that there might be a fight. He did chase Genius, and Genius did indeed stay away for a few days. But he returned, much to my relief. There were other returnees as well: three smallish squirrels I suspected were the gang of three I had banished the previous fall, and whom I further suspected might be Genius's offspring. That gang included the little male who was so tough. Genius appeared to tolerate them. He arrived shortly after they did one morning, and allowed one of them to occupy his spot on the sill while he moved over to Yoko's window. I then began to worry about the effect of the tough little son on Genius's psyche. On May 13, I recorded that Genius, coming in soon after the little ones, looked huge. That date is the last time that year there is a record of my having seen Genius. For the rest of the month, the entries in the squirrel diary say, "No Genius today," or express concern that I haven't seen him, or state that I miss him.

On June 7, we left for Colorado and the Aspen Center for Physics, where Larry had worked for a number of summers. For the preceding few days, Jeannie had been putting out nuts on her windowsill. I was still feeding anyone who showed up, to let all know there was food in both places. I left Jeannie with a two-month supply of nuts, ten dollars for more if necessary, a bag of acorns, and a complete description of Genius. I prayed that the thirteenth might not prove an unlucky day.

On June 10, in Aspen, we got a card from Jeannie.

Dear Grace and Larry,

Genius rides again! I opened the window this A.M. and guess who appeared. I think he "knew" it was about time for you to be going away and he thought he'd better check in. He ate eight hazelnuts and all the peanuts! Probably to convey the impression that he is a dependent squirrel. He didn't look like he needed any food.

Tra la,
Jeannie

Jeannie's note was a kind of coda. Genius was nowhere to be seen when we returned in August. But at least I was able to enjoy Aspen.

14. SOCIAL WORK

I HAD NEVER CONSIDERED Darwin to be a great scientist. Evolution seemed obvious, hardly something anyone would have to "discover." But the more I discovered how little was known at the time Darwin formulated his theory, the more I respected him as a scientist. I knew little about him as a person. In recent years, mostly because of that fine series on Darwin that Public Television kept showing and reshowing, which prompted me to do some reading on him, I came to like him enormously. Darwin's argument against the existence of God, based on the suffering of animals, was what finally hooked me on him.

In his autobiography, Darwin says he had been quite orthodox while on the *Beagle,* and would quote the Bible as authority on some points. He became less of a believer because he was troubled by the miracles. (I could relate to that. When I first began to read, at six, someone gave me a book of Bible stories. I don't know what the intended effect was—I no longer remember who gave me the book—but the actual effect was to make me a kind of atheist: Since I had not seen God as a pillar of fire, I didn't believe *anyone* had.) Darwin goes on to say that if Christianity were true, non-believers like his father, his brother, and some of his best friends would be everlastingly punished, which he found impossible to accept. But the clincher for him—after a number of other arguments—was the suffering of animals. Darwin said he might be able to understand the suffering of humans—some said it served for their moral improvement—but "what advantage can there be in the sufferings of millions of the lower animals throughout almost endless time?" Moral improvement aside, I could see that for people suffering in this world might have its compensation in the next. But animals aren't allowed into heaven because, presumably, they have no souls. All nonbelievers have their reasons. I like Darwin's best.

In addition to the major social work projects like the one for Runty, there were minor problems from time to time. Around the Christmas just before Genius's ear was split, for example, I shelled nuts for a very gentle male who had difficulty on his own. He had a patch of fur missing from his forehead and one eye that wouldn't open. He took half an hour to shell even a peanut. When he learned that I provided assistance, he began coming regularly and eating inside on the windowsill. In a month, his eye was open and he was eating peanuts at a respectable rate. When I handed him a filbert, however, he would drop it deliberately. He knew his limitations, but was not very bright in other respects. He came in over the top of the screen one day and then couldn't find his way back out. He kept standing near the exit, which wasn't open. But he showed no fear or panic, even after falling off the sill. I continued to crack hard-shelled nuts for him, and he waited patiently for them. He appeared to use his tongue to help him chew; it thrashed about visibly. He had teeth, I knew, for when he took a nut from my hand, I could feel the lower ones. I named him Perry, for his possible periodontal disease. Perry evidently did not feel the need for extended welfare, just temporary assistance. After several weeks, he ceased coming.

There were several periods when Larry and I were in need of

treatment ourselves. We never got rabies or tetanus, or even badly bitten—but we did get fleas. We had volcano-shaped, isolated red bumps below our ankles, which lasted for weeks, especially since we couldn't resist scratching. One, on my second toe, didn't bother me much during the day, but the instant I took off my shoes, the bump itched maddeningly. When the red bumps started to appear higher up—Larry got one on his hip—I took action: flea soap, powder, washing everything we contacted, and getting rid of my fuzzy red dynel slippers, probably the fleas' main hideout. Which action was most effective I can't say, but no new bumps appeared and the old ones faded in a couple of weeks.

Happy squirrels, to paraphrase Tolstoi, are all alike, but every unhappy squirrel is unhappy in its own way. Unfortunate would be a better word than unhappy, since the squirrel may not be aware of being unfortunate, and therefore not unhappy. At the beginning of February, I had another client, one with a more serious handicap. His left front paw had no digits—only a bit of pinky. His paw was chopped off at the wrist. I couldn't tell whether his deformity was congenital or not. The fur on his stump looked normal. Out of habit, I noted that he had a notch and scar on his left ear, a smaller notch on his right ear, and a tiny nick in the bottom of his right eye. But it was his paw that set him apart. He was quite relaxed, took nuts from my hand while Larry and a student worked in the room. After a while, though, his sawing of filberts got slower and slower, with longer intervals of clicking, to sharpen his teeth, in between. I cracked a few of the filberts for him. Seeming grateful, he moved closer to me on the sill. A squirrel tried to come in while he sat there, Round-the-Grill, I think. He routed her. I named him Philip—our one exception to the no-proper-names rule—after the clubfooted protagonist of Maugham's *Of Human Bondage*.

Philip was a game little fellow. It must have been difficult for him to get up to our apartment, yet he came regularly. The second time in, he turned down a peanut and an acorn, took a filbert from my hand, and went at it. His zest soon petered out, though. I cracked one and held it out to him. It must have been too far away, for he grabbed my finger with his good paw and pulled my hand toward him. He had a surprisingly strong grip, a compensation for his club paw, I supposed.

He soon settled into a routine, coming later than the others, toward midmorning. He selected as his dining area the entrance itself, just under the window sash, and strongly resisted my attempt to evict him when I had to leave. A week after his first visit, I saw a scar along the length of his back. It started at the top of his head, between his ears, where it was almost the full width of his head. The fur was missing there, and small scabs dotted his pink skin. As the line of missing fur continued down his back for about six inches, it narrowed and contained no scabs. The long pink line was not a scar in the skin but a scar in the fur. I had not seen it before. It was so noticeable, though, that I was certain I could not have overlooked it, no matter how engrossed I was in Philip's clubfoot. The scabs suggested that the scar was squirrel-made, the product of a fight. I suppose they could also have come from the scratching of an itch, if he had a skin problem. But a skin problem would not account for the unscabbed line of missing fur down his back.

Philip would gallantly take a filbert from the cup and begin to saw. Always, the vigor of the sawing would diminish audibly after about thirty seconds. The operation then became torture—for me, that is. Finally, when I could stand it no longer, I would shell a filbert for him. But his pride—or his greed—made him continue a while longer with the one in hand until eventually he would give up and accept my offer. He seized almonds, managing them ably, but would not consider acorns or peanuts. He would wait while I cracked filberts for him, occasionally clicking with impatience. (These small creatures have large frustrations.) Larry sometimes clicked, as well; he objected to my shelling nuts, for fear Philip would become dependent upon me. But I believed that I was building up Philip's strength, so that he would be able to handle filberts himself. I didn't know why he wasn't able to shell them. He appeared to be full grown. He was able, despite his clubfoot, to get to our apartment and to shell almonds. His sawing indicated that he had the teeth for it. My fingers indicated the same. Philip, Perry before him, and the female Curly Ears, all seemed to have dental problems at about the same time. (I point to the fact that both genders had the same problems to forestall suspicions that they were associated with gender-related activities such as fighting or nest-building.) I wondered if a kind of "tooth flu" was going around.

Philip had a healthy appetite; he consumed about ten almonds and ten filberts at a sitting. He either held, or was close to holding, the eating record. I suspected that he loaded up for several days, for he did not come every day, not while I was there to observe, at any rate. He usually left with an almond.

While cracking filberts for Philip one morning, I was struck by the ridiculousness of the situation: I, a human, cracking nuts for a squirrel. Then the Jewish mother aspect struck me: I was serving more than three different main courses every day—filberts, almonds, acorns, peanuts—to satisfy the different needs and preferences of my different charges, like Larry's mother at Friday night meals. And I was even getting complaints: I had gone to Sloan's for peanuts and found that they were out of them; once there, however, I bought Sloan's rice pudding for Larry. "I get rice pudding only when they need peanuts," Larry whined.

One of the filberts I cracked broke into pieces. In attempting to take one of the pieces, Philip bit my finger. It didn't hurt—Philip was quite gentle—but I evidently pulled back, for Philip was pulled off the windowsill. He hung from the edge by his good front paw and one back paw, spread out like a bat. I scooped him up onto the sill. He was soft and light, so much lighter than his strong one-pawed grip had led me to expect. He showed no fear or annoyance that I had handled him. He continued to eat the fragment of filbert—which had not fallen from his mouth when he fell from the sill—and then looked up for more. In general, the squirrels were not frightened when they slipped or fell or their footing gave way, as when one stepped on a magazine that extended out over the edge of the desk. Falling is apparently all in a day's work.

While Philip was unmistakable, I did have some difficulty recognizing some of my other clients. In the late afternoon near the end of February, for example, a squirrel came in, fur missing from his face and a notch on his ear that looked familiar. He obviously knew the setup at our place, for he went straight to the cup. He took filberts, dropped them, did the same with almonds and acorns. What did he want? Peanuts? That would have been most unusual. He clicked at my lack of understanding. Then I realized it was Perry. I hadn't seen him in weeks. I cracked some filberts for him, which he accepted eagerly, and gave him some soybeans, which he accepted less eagerly. He looked at me intently as he chewed.

My failure to recognize Perry was not entirely my fault. As the weather turned warmer—60 degrees, though it was still February —visits fell off. There was also the possibility that the squirrels were receiving assistance elsewhere. Jeannie's colleague in 5 Washington Square North supplied peanuts and Oreo cookies, I was informed. I had meant to interrogate her about the extent of her benefits, as they may have been at the root of the dental problems I'd seen.

By the end of the second month of Philip's visits, he was coming early and making a racket on the window filter so that I would let him in. I would hand him an intact filbert to keep him occupied, so that he wouldn't click at me while I cracked a few for him. If, after about five minutes, he had had no success, I would hand him one I had just cracked. He knew from experience that this filbert was cracked—I had become so skillful that the nuts appeared whole, often having only a hairline crack—and grabbed it, dropping the one he had been working on. He would search diligently for the crack he knew was there, turning and turning the nut until he found it. After consuming the cracked nut, he would return to the uncracked one, and often succeeded in opening it, as if the cracked one had given him the strength to handle the uncracked one. On those occasions when he failed to find the crack, he would give up and start to saw the shell. He would eventually hit upon the crack and the nut would be shelled much more rapidly than if he had started from scratch. I liked to think he was under the impression he had done it entirely on his own, and that I was instilling confidence in him.

Philip was absent for more than a week at the beginning of April. When he returned, it seemed to me his paw was regenerating digits. There were little bumps where, before, there had been the one large bump that constituted his paw. And the pinky appeared bigger. Since this was obviously impossible, I assume that Philip had lost some winter fur, which had prevented my seeing the little stumps that had been there all along.

Philip ceased coming as suddenly as he had appeared. From the beginning of February to the beginning of April, he had come at least every few days. Then, one visit after his week's vacation, and no more. I missed the little fellow.

15. SWEETIE

ONE OF THE SWEETEST squirrels we had dealings with was Sweetie Longtail, who is now probably dead. He bore that silly name because it came in two installments. He was simply Longtail for a time, until we got to know him well.

Our introduction to Sweetie was actually unpleasant. He was part of the band of marauders, those teen-age hoodlums, that had appeared the previous year, and although he himself was well behaved, he was guilty by association.

He turned out to have a special sweetness that set him apart from your common-or-garden gentle squirrel, and it was not just the contrast with his disagreeable sister. He reminded me a bit of Notchko. But while Notchko's was a mature, dignified, confident gentleness, Sweetie Longtail's was a youthful, shy gentleness. Sometimes we would be at the kitchen table and would suddenly notice him standing up tall, paws to chest, on the other side of the threshold, looking at us, as if waiting to be invited in. We *would* invite him in, of course, and offer him something to eat. He was one of those "touchers" who place a paw on your hand as they take a nut from your fingers with their teeth. (Except in frigid weather, the paw feels delightfully warm.) I had wondered whether the

113

touching might not be part of an effort not to bite a finger when taking a nut. As both nut and finger are right in front then, where squirrels don't see well, touching might help in separating finger from nut. But Sweetie Longtail's touch occurred an instant *after* he took the nut, and since it was too gentle to serve as a push-off to change direction, I saw it as an acknowledgment, a kind of thank you.

As sweet and gentle as Sweetie Longtail was, however, he browbeat his sister Halftail. She became the fastest eater of all the squirrels. She had to get *something* inside her before being chased off the premises. She became an eclectic eater as well: peanuts, soybeans, cantaloupe seeds, even watermelon seeds—*anything* as long as it was food and close by. (Those seeds, incidentally, required considerable preparation to avoid attracting insects along with squirrels. Washing cantaloupe out of its seeds is a sticky business.) Halftail and Sweetie Longtail generally came to the house together for their morning meal.

Halftail became more and more agile, and adept at side-stepping or jumping over Sweetie Longtail. She developed a technique that Curly Ears and Round-the-Grill had worked out earlier: She'd climb up the burglar grill to the top of the lower half of the window and eat there. I used to hand nuts to Curly Ears and Round-the-Grill through the slats of the blind. Now I did it for Halftail, while Sweetie Longtail looked up, bewildered, a shell having hit him on the head. He was not actually cruel to Halftail. He merely made the requisite threatening lunges when he found her near the cup, and she responded as though he truly meant business.

When we returned from Aspen in late July, the summer Genius disappeared, for the first few days our only visitor was Longtail. Then Halftail showed up. She was as grabby as ever, attacking me when I tried to add to the supply of cantaloupe seeds she was devouring. Longtail's tail seemed longer than ever, and he seemed especially sweet. That was when Sweetie was added to his name. He walked across me, as I sat on the chair near the window, on his way to the kitchen to try to bury. He readily obeyed the "No!" that went with attempts to bury in plants. Spying a bra hanging by a shoulder strap from the doorknob, he pulled it to him for examination. That done, he stood up tall and looked at me, his way of asking for nuts. Halftail would click with anger or annoyance when she had to wait for nuts.

I figured that Sweetie had a high IQ. He had been at the cup picking out almonds from among the acorns and, in the course of selecting, dropped an almond, apparently deliberately, onto the base of the rubber plant. When Sweetie took the last almond in the cup and went out with it, I retrieved the one he'd dropped on the plant and put it in the cup. Sweetie returned, took that almond from the cup, but before leaving with it, he jumped down onto the plant. Was it to check whether he was right in remembering that there had been no almonds left in the cup? He did not ordinarily jump onto the plant.

On another occasion, he again exhibited what appeared to be memory intelligence. Larry was seated with one leg crossed over the other and had decided to make Sweetie come up onto his upper foot to get a nut. Sweetie obliged. Then Larry got him to shinny up his leg to get onto his lap for a nut. The two of them kept up the game for a good part of the afternoon. I had the impression Sweetie did not object, and rather liked the interaction. Two days later, the instant I opened the screen, Sweetie entered, plopped directly down to the floor, and headed for Larry's foot.

Why shouldn't Sweetie be smart? He was probably Genius's son. He had inherited Genius's brains. Halftail had inherited his temperament and his tail.

* * *

After some months of relatively uneventful relations with Sweetie, we began to notice in him a growing peculiarity. He would go to the cup, passing a line of peanuts on the way, take a filbert from the cup, but continue to look in and to rummage a bit. Then he would place the filbert next to the cup and select another. The second one, too, he would place beside the cup, and go on rummaging. He might take one of the two he had placed next to the cup and return it, or he might pick it up and leave with it, putting his nose against each peanut he passed on the way out. Sometimes he would exchange the filbert he was carrying for a peanut.

We had long since ceased to pay attention to squirrels at the cup except to notice vaguely when the nut supply was getting low. But Sweetie's behavior was hard to ignore, what with nuts rattling in the cup, and filberts being dropped to the floor. Larry and I became conscious of Sweetie's erratic behavior at about the same time. Larry couldn't stand it. He finds indecision in people annoying; in a squirrel, he found it maddening. Sweetie would sometimes go through eighteen nuts before making his final choice. Larry couldn't concentrate. Against his will, he would watch Sweetie's operations. The elaborate selecting went on for weeks. Was Sweetie searching for the perfect filbert?

It was by accident that I learned what was troubling Sweetie. Since he was such a nice little guy, I decided one day to give him a walnut, even though it wasn't anyone's birthday. But since I didn't want him to bury it, I cracked it and gave him half. He took it instantly and devoured it. Likewise the other half. I gave him a partially open one for the road, but I could hear him outside the window opening it the rest of the way and eating it. He took another half walnut away with him. The next day, he was back to choosing, but took nothing. Larry firmly believed that we should not spoil him with more walnuts. When Sweetie took a peanut from the cup and, because it was raining, ate it inside, it struck me that prior to yesterday I had never seen Sweetie eat. It also struck me that he had been taking peanuts most recently. Then I realized, finally, that it was not walnuts Sweetie wanted, but nuts that were cracked, or crackable. To test this hypothesis, I shelled a filbert and held the meat out to him. Not an instant's hesitation. I shelled another. Again, he took it immediately. The same was true for two almonds and some more filberts. He was so stuffed, after a while, that he started to bury a shelled filbert in Larry's slipper—with Larry's foot inside. Larry drew back, naturally, and Sweetie made for a plant. I frustrated him there, and at the next plant he tried. Sweetie gave up and ate the shelled filbert. I gave him an intact one to take away with him. The next time he was in for a meal, he would take nothing with an intact shell: Sweetie wanted his nuts cracked. Why, I did not know.

A new routine developed. When Sweetie entered, he no longer went directly to the cup at the far end of the windowsill, but immediately jumped to the desk, where he had received his first shelled nut from me, and waited. I would shell some filberts for him and hand him one; the rest I intended to place on an aluminum ashtray next to him. Sweetie was not a rapid eater; it would sometimes take half an hour for him to feel sufficiently satisfied to go to the cup for an unshelled filbert for the road. Fortunately, he was also not a big eater, for instead of leaving nutmeats on the ashtray, I would personally hand him a filbert, then a half-walnut-on-the-shell, then another filbert, then an almond—for a balanced diet.

This procedure continued for weeks. On the days when my teaching schedule prevented English breakfasts for Sweetie, he probably made do with peanuts. There was no point in my shelling filberts and leaving them outside; chances were they would not be eaten by Sweetie. Squirrels are as lazy as people when given the chance, and won't waste time and effort shelling nuts when they don't have to. For some squirrels, especially females like Halftail who had to consume as much as possible before being chased from the premises, shelled nuts were a godsend. I made a point of leaving peanuts outside, so that Sweetie would get *something* at least. No one other than a stranger would condescend to eat peanuts at this stage.

Sweetie did not simply resign himself to dependency; he did make some effort to avoid it. From time to time, he would take a whole filbert outside on the fire escape and set to work on it. Unsuccessful, he would bring the filbert back and trade it for a peanut on the sill, or return it to me directly, to exchange it for a shelled filbert. I wondered whether his bringing back the unshelled filbert was his sign to me that he wanted cracking or whether it was simply that, for a squirrel, a nut in the mouth is worth two in the bush, even if it can't be eaten.

One day, I noticed that Sweetie had awfully large lower teeth. I had never before been conscious of lower teeth—large or small —I simply knew they played a part in shelling. I think the lower teeth do the initial sawing of a ridge in a filbert shell into which an upper incisor is then pressed to split the shell open. It was easy not only to see Sweetie's lower teeth but to feel them when I handed him a nut fragment. The teeth felt large and dull, like the single tooth of a human. In fact, I thought at first that there was only one lower tooth in front. Only when I could see Sweetie's teeth even when he was not eating did I realize that there were two teeth.

Sweetie had buck teeth. But buck lowers rather than uppers. I kept trying to put my finger in his mouth to feel what his upper teeth were like. I couldn't see his uppers no matter how I bent and contorted myself in front of him as he sat on the desk. He went right on eating, giving no indication that he found my gyrations in any way peculiar.

Sweetie's buck teeth continued to grow. When I told my

friend Shirley Noakes about Sweetie Longtail, she started to refer to him as Sweetie Longtooth. Sweetie's teeth were not a joking matter, however. His left cheek became swollen, so much so that I could identify him by it.

Once again, I got on the telephone. I called the zoo, the Museum of Natural History, the ASPCA, animal hospitals, and veterinarians I found in the yellow pages. I no longer remember who said what, but a general pattern emerged: Sweetie's lower teeth would continue to grow, penetrate his brain case, and kill him. Why this was occurring could not be determined without examination. Dr. Hunziger, an exotic animal specialist someone had recommended, said Sweetie might have malocclusion, and if his lower teeth were pulled, the upper teeth would have nothing to work against and he would be in the same shape—or worse—than he was in now. He suggested having the lower teeth filed down to proper size, which might give him six months of normal eating.

Most of the doctors and hospitals I called refused to handle Sweetie. He was an exotic—which is how I got onto Dr. Hunziger —while they treated only ordinaries. My periodontist, Dr. Froum, offered to make Sweetie a bridge, tying the two large teeth—after filing—to neighboring teeth to prevent their growing to outsize proportions again. (Would that make Froum a rodentist?) All remedies involved trapping Sweetie.

I called Claus Kallmann, my thesis advisor's son, because I had been told that Claus had a trap and because Claus is a geneticist and Larry had been badgering me about Sweetie transmitting his defects to his offspring and our having to shell filberts for a whole colony of toothless squirrels who would be our responsibility because we had brought them into the world, so to speak. Claus's guess was that Sweetie had had an accident and therefore was not likely to transmit his dental problems to offspring. I was not really concerned about transmission, I asked only to placate Larry, who didn't like to feel he was interfering with nature. What worried me was the damage to Sweetie's psyche that trapping might cause. Claus pooh-poohed my fear of psychological damage, and felt we should trap Sweetie before his wound became infected. He assured me that he trapped squirrels regularly to remove them from his property, and if he didn't transport them far enough way, they returned almost immediately—which hardly indicated trauma. It entered my mind that Claus might have the indifference that comes with working with animals regularly, especially since those he worked with were fish, which had no real psyches to speak of. Nevertheless, I acted on his recommendation. I ended up borrowing a trap from Erwin Cohen, whom I also knew from my days in Professor Kallmann's lab. Erwin brought it into his office at Academic Press and I went there to get it. Erwin had put it in a shopping bag, the handles of which he had extended with rope, to facilitate my carrying it the nine blocks from Academic Press to my house.

The trap was about three feet long, one foot high, and one foot wide, made of wire mesh somewhat like the chicken wire I had used to protect the plants. There was a door at each end that hooked to the top, so that when the doors were open the apparatus looked like a two-door garage—doors front and back—with the doors up. In the center of the trap was a metal platform for nuts. A squirrel, entering the trap to get a nut on the platform, would presumably have to place a paw on the platform, thereby releasing the doors, to which the nut platform was attached. The squirrel would be inside, unharmed, disturbed only by the bang of the doors and loss of freedom. The trap bore the trade name *Hav-A-Heart.* I removed the trap from the shopping bag and set it on the far end of the desk. I wanted it to become an accepted piece of furniture.

Meanwhile, Sweetie was as active and cheerful as ever, despite an increasingly swollen face. I was feeding him half-walnuts, which he was able to get out of the shell and which I thought gave him the feeling he was doing something for himself.

At night, I practiced using the trap. Practice was necessary because either the Hav-A-Heart people had too much heart and didn't trap, or this particular trap was old and worn and didn't work properly. I pushed on the platform with more force than what I estimated to be a one-paw lean by Sweetie and nothing happened. The door supports were bent and didn't release. What I was practicing, therefore, was releasing the doors from outside, so that I wouldn't have to rely on Sweetie doing it from inside.

I had made a tentative appointment for Sweetie at the Washington Square Hospital for Animals. Dr. Rosenfeld would not make a commitment over the phone, but had agreed to look at Sweetie and to do what he could. In addition to Dr. Rosenfeld's sympathetically discussing the problem, his hospital had the advantage of being located a block and a half from my house, making it unnecessary to add a New York City subway or taxi ride to Sweetie's distress. The appointment was tentative because I could not be absolutely certain Sweetie would show up on any given morning. Patients were admitted to the hospital only from ten A.M. to noon and four P.M. to six P.M. Monday, Tuesday, Thursday, and Friday. Sweetie was showing up at all hours. If he appeared after noon, the plan I had could not be carried out. I wanted him in early, attended to, and out before evening. Otherwise, there would be a problem getting him home. I did not know where he lived.

I was getting nervous—I am not a nervous type, but Sweetie was making me experience feelings I hadn't had since the last time I took exams. I was torn between wanting to help him and worrying about interfering in the domestic affairs of a wild animal. Probably the main source of nervousness was my fear of losing him —I was afraid he'd hate me for trapping him and would never darken our fire escape again.

I psyched myself for the ordeal. Wednesday was no good, the hospital was closed. I decided to try Thursday. But the Wednesday on which I made the decision was July 13, 1977, the date of

New York City's second blackout. Electrical power went off at nine-thirty Wednesday night and by quarter to ten Thursday morning, when Sweetie showed up for his breakfast, the power was still off. The doctor would not be available, or, if he were, he would have no light with which to perform an operation. I must admit I was relieved. I was undoubtedly one of a very small number of New Yorkers to whom the blackout provided comfort.

Friday was hot, and this generally keeps squirrels away. A decision on trapping might be made for me. It was nearly ten o'clock before any squirrel appeared. It was the new little male I had not yet named. The yellowish Bambi-like outlines of his eyes and his consumption of canteloupe seeds proclaimed his youth. He was just making the connection between me and the loot he found on the windowsill, and realizing that it might be intended for him. Until now, he had acted as though he were stealing. I put out more seeds. He tried to take some that were stuck to my hand. I felt his sharp teeth and pulled back. He was pulled off the sill and dropped to the floor. I was not hurt, and he was not frightened, but both of us were a bit ruffled. He went back to the seeds, ate all of them, and left with a peanut I offered him.

Fifteen minutes later, someone was on the desk taking a half-walnut from the aluminum tray. My heart pounded. Then, because the walnut was being eaten so quickly, I knew without looking at his face that it wasn't Sweetie. It was the little male; he had noticed the walnut earlier and had come back for it.

Sweetie didn't appear until amost dark. He looked awful. His face was so swollen that in profile he looked like a miniature rhino. My hesitation about trapping him disappeared. I shelled a walnut for him. He took a piece outside to eat. I put the rest of the walnut, two halves of a second one, and a cherry on his tray near the trap. He returned for one of the walnut halves and left for the evening.

Sunday morning, Larry announced, "Sweetie has water on his face." Opening the screen to let him in, I saw a black string about three-quarters of an inch long hanging from Sweetie's bad cheek; it was what Larry had thought was water. His swelling was down and he no longer looked like a rhino. He came in and stared at the venetian blind, which was much lower than usual; then he stood up and swatted the cord with his paw. He couldn't feel too sick, I thought. I gave him half a walnut, which he carried out. I put food into the trap, to get him used to taking nuts from there. He walked right into the trap on his return, but somehow missed the bait, walked out, and climbed onto the top of the trap. As I handed him a piece of walnut, I tried to knock the string off his cheek. It was pus. I felt relief. Maybe the infection had cured itself and I might not have to trap him.

Soon the little male arrived. Since the cup was empty, I handed him a filbert from the trap. He left with it. Five minutes later, he was back, and grabbed another filbert from the trap, and now, with a little cry, was running from Sweetie, who was on the fire escape in front of the exit, a half-walnut shell in his mouth. Even in affliction, Sweetie was asserting authority.

Sweetie came in several times for bits of walnut and a shelled Brazil nut I held in my hand. Seeing a half-walnut further back in my hand, he exchanged the Brazil nut for it. Finally, I handed him a filbert with shell intact, which he took, undoubtedly for the road—that is, to bury, commanded by his genes, for there wasn't a chance in the world that he would be able to crack it open.

The little male returned after about twenty minutes, crying. Crying is the only word I could find for his gibbering and his *waah waah* that ended in a high pitch. His crying went on and on *as he ate.* It did not seem to be directed at anyone. Gradually, it subsided, and stopped altogether when I handed him a shelled Brazil nut. He had by then eaten peanuts, canteloupe seeds, and a cherry (which he did not peel). I never did learn what he had been crying about.

At night, I practiced springing the trap with everything in position. A dress rehearsal.

T-DAY

Monday, July 18,—T-Day—the alarm went off at 8:10. Larry had been up since seven. Hearing the alarm, he entered the bedroom, saw a squirrel outside, and started to whistle at him. "Shhhhhh!" I hissed, "It's *him!* He'll have to spend an hour in the trap if we let him in now!" We played dead.

It seemed like several minutes that the three of us, Larry, Sweetie, and I, stood motionless, each waiting for movement on the other side of the screen. Sweetie went away; Larry prepared breakfast while I washed and dressed in order to be ready.

At eight-thirty, I slid the screen open. The little male! He might louse things up. I put some acorns outside for him. He wanted no seeds; he'd apparently had his fill of them over the weekend. Soon, from his nervousness, I could tell that another squirrel was approaching. Sweetie. His face looked better. There was no long thread of pus and the swelling was down. I handed him a piece of walnut, which he took outside to eat. The next piece he would have to take from his ashtray, which I had placed in the trap. I prepared to spring the doors. Sweetie was suspicious. He wouldn't go in. He must have sensed my nervousness. My hands were shaking. I had hardly slept. The trapping of a living thing had upset me more than I would have thought possible. Sweetie jumped onto the back of the desk chair, compelling me to hand him a piece of walnut, which he again carried outside.

When he came in again, I showed him a half-walnut inside the trap. He was still suspicious or nervous and wouldn't go inside; he climbed on top of the trap. I was afraid he might spring it from above, so I gave him the half-walnut in order to make him get off the trap and leave. I feared the whole enterprise would collapse.

It was clear that Sweetie had to be trapped. His cheek would swell again with a worse infection as his teeth continued to grow. It was nine o'clock. I phoned the hospital to ask if Sweetie could be admitted that morning. I also moved the chair away from the desk. Sweetie came back and walked into the trap. I banged the doors down. My heart banged as loudly as the doors.

I lifted the trap and, holding it in my arms, wrapped some pages of *The New York Times* around it so that dogs wouldn't see what I was carrying, then raced out of the house. Sweetie was thrashing around inside and trying to bite the wire of the cage. His long teeth caught on a section of wire and he let out a little cry. I pushed the teeth down and back from around the wire, freeing them. They were more than half an inch long. I kept talking to Sweetie as I hurried along the street, telling him everything would be all right. I wanted him to hear a familiar voice and I wanted to tell myself that everything would be all right.

Only one dog was in the waiting room when we entered the hospital. I asked the receptionist if Sweetie could be given a tranquilizer. The doctor, who had just then come out of an inner room, heard me and said, somewhat impatiently, that he would be only a few minutes. He ushered the dog into the inner room.

Three more patients came in with their attendants. Dogs. They had appointments. When I explained about Sweetie, one woman relinquished her dog's place to him. The young receptionist said Sweetie was next anyway. After about ten minutes, the doctor came out, lifted Sweetie's cage off my lap by the handle on top, and carried it into the inner room. I followed. He set the cage down on a table, took up a hypodermic needle, and was about to administer the tranquilizer when Sweetie noticed the half-walnut in the cage. Sweetie picked up the walnut and began to eat. The doctor put away the hypodermic and said, "Sweetie doesn't need a tranquilizer, you do. He's relaxed. I'll put him in an anesthesia

chamber when I have the time to watch the effect of the anesthetic—I don't know the proper dose for a squirrel. Come back
later in the day." Sweetie's infected cheek was bloody from his
having scraped it against the wire of the cage. I asked if I could
stay with him until the anesthetic was administered. "I can't work
with Mamma hovering over me," the doctor replied. "Come back
at five-thirty."

The receptionist said to leave twenty dollars and fill out an
admission form.

The first line of the form stated:

> This is to certify that the undersigned is the owner
> and/or authorized custodian of the pet animal described as
> follows:_____.

I filled in the blank with "squirrel." In the next blank, after:

> It is understood that the above animal has been left in
> the hospital for the purpose of _____

I wrote "dentistry." I signed the blank over "Owner/Author-
ized Custodian" and left.

Outside, in the street, I feared I had been too flippant. I
returned to write in specific instructions about filing Sweetie's
teeth. The receptionist, who looked about fourteen years old,
refused. "Why?" I asked. "Because it's silly," she said. Just then
Dr. Rosenfeld came into the waiting room. I told him what I
wanted to write on the form. He said, "I know what to do. I
remember our phone conversation."

At two P.M., I telephoned to ask whether I could pick Sweetie
up early, if he was ready. Dr. Rosenfeld answered and said Sweetie
was still unconscious and to call for him about five o'clock.

I was at the hospital on the dot of five. Sweetie was fast asleep
in his cage. I carried the cage home as gently as I could and set
it down on the desk exactly where it had been when Sweetie was

trapped. I opened the doors and placed the aluminum ashtray, with bits of walnuts and two cherries on it, in front of Sweetie, in the hope that when he awoke he would think the entire episode had been a bad dream. Sweetie stirred every now and then; his eyes were not entirely closed.

Dr. Rosenfeld had filed Sweetie's lower teeth down to a little below normal. He said Sweetie had no upper teeth at all; he could not tell whether that was the result of an accident or whether Sweetie had been born that way. When I asked why Sweetie didn't file his lower teeth on a wire or a branch, Dr. Rosenfeld explained, "Because he's never been to school. He doesn't realize the abrasive nature of things. If squirrels had been born without upper teeth, there would be an instructive mechanism for sharpening the lower teeth. Birds know instinctively how to sharpen their beaks."

It was past six before Sweetie awoke. He seemed to be licking the inside of his mouth. After about a minute, he tottered out of the cage, jumped from the desk to the windowsill, and tried to leave. I had closed the window. He was in no condition to go home alone. He fell asleep in front of where he would normally exit, his nose on the sill. I put a walnut and a cherry in front of him. He roused, tried to grasp the cherry, and fell asleep again, his nose on the cherry. He kept waking, attempting to leave, and falling asleep again. He was not interested in eating. At seven P.M., he woke and tried to get out by climbing up the window grill—something he had never done. He reached the ledge that is the top of the lower part of the window and went to sleep there. At 7:50 he stirred, changed position, scratched, and went back to sleep. Ten minutes later, he made motions with his paws as if he were rolling nuts, the same motions he had made earlier when he tried to grasp the cherry.

It was getting quite dark outside. Jeannie phoned and suggested we keep Sweetie overnight in the cage. She thought we could get him back in the cage by catching him in a towel, a technique she used with Mouse when he acted ornery.

At 8:35, Sweetie opened his eyes. He climbed down from the ledge and tried to get out the exit. I had shut it since I wanted him to be completely alert before he left. He jumped to the desk, investigated the cage—he seemed to remember it—jumped down onto the base of the rubber plant, onto the floor, up onto the bed, back down to the floor, up onto the bookcase, over books and the television set, and then—I had opened the exit now—out. Larry ran downstairs to see if he had made it down all right. He couldn't see Sweetie on any of the fire escapes or in the garden. We were relieved to remember that even while groggy Sweetie was able to climb and jump with agility from the desk to the windowsill through the leaves of the rubber plant.

We were to leave early the next morning for three weeks in Europe. Larry was to attend a conference in Paris for one week. The rest would be vacation. Our departure was one more reason for the urgency about Sweetie's operation.

16. POSTOPERATIVE TREATMENT

I THOUGHT ABOUT SWEETIE many times during our three weeks abroad. I was relieved that the operation was over, but I was anxious about our future relations.

On our first morning back, someone was outside the window at nine-thirty. I opened the screen and Halftail came in. She took a peanut from my hand, giving me a gentle touch as she did so, as if in greeting. She ate several more peanuts that I put on the windowsill. Later, when I saw peanuts being selected there, I knew it was Sweetie. He jumped onto the desk, where I plied him with walnuts. No peanuts for my friend Sweetie! He appeared to be fine, but I could see his lower teeth. I had expected them still to be shorter than normal after only three weeks. Sweetie acted as if we had never left and as if there were no such thing as a trap.

At night I spoke to Jeannie, who had taken on her customary job as squirrel sitter while we were away. She reported that Sweetie had knocked at her window the morning after his operation. By knocking, she meant that he and Halftail put their paws against the glass of the fire-escape window, which she kept closed because of Mouse, and she could hear the tap of their nails. She would respond by opening the window and putting nuts on the sill outside. I am certain that Sweetie and Halftail knew their nails made a sound that called forth a response in Jeannie—they made no noise on our screen when the hour was earlier than opening time and they would get no such response from us. If it was late, however, or if they saw us moving about the room, they would climb onto the screen and make a racket. Jeannie also reported that during our absence Sweetie had been shelling peanuts, acorns, and some almonds.

Things went along relatively placidly for a while, but I was anxiously watching Sweetie's teeth. I knew I would have to take him in for another operation, but I didn't know when. We were to leave in mid-September for a year's sabbatical at Harvard. Larry would do research at the Harvard Smithsonian Center for Astrophysics and I was to be an honorary associate of the Nieman Foundation for Journalism. We planned to come in to New York every three weeks or so to visit Larry's father, who was old and housebound. Sweetie's operation would have to be fitted in with these visits, which would not be easy. He could only be admitted to the hospital during the week, and there would be little time to get him used to the presence of the trap. Since he would have become accustomed to dining at Jeannie's, it might even prove difficult to get him into our apartment.

I was becoming thoroughly acquainted with Sweetie's habits during the weeks before our departure. He would come in in the morning at opening time and ignore shelled walnuts. He would elaborately select from among the peanuts, filberts, Brazil nuts, and acorns those with the sturdiest shells. He was obviously burying. After about an hour of selecting, he had worked up enough of an appetite to eat two entire shelled walnuts, scrape an acorn, and open a peanut by himself, with time out to chase Halftail away from the watermelon seeds I had put out for her. With a little cry, she would run, and Sweetie would try a couple of watermelon seeds. They required more teeth than he had, however, and so he'd take up a peanut. When he refused any further shelled walnuts, I knew he wanted a filbert in an intact shell for the road. After his departure, Halftail would return, to finish his leftovers.

Sweetie's lower teeth were getting longer, and Sweetie himself was getting spoiled. I claimed full responsibility for both phenomena. With all the shelled walnuts he was getting, he was not filing his teeth at all, and at the same time he was burying filberts that other squirrels would gladly have eaten. And now, at the end of the summer, the supermarkets were out of filberts. Gristedes did not expect any until the new crop would arrive in about a month. I started supplying Brazil nuts for burying in order to make my supply of filberts last longer.

For his own good, I resolved to be unkind to Sweetie. No more shelled walnuts. He would get almonds and shell them himself. Easier said than done. Sweetie didn't want almonds—not unshelled, not half-shelled, not even shelled. I informed him that that was all he would get. Disgusted—what other way can I describe his manner?—he left me standing in the bedroom and headed for the kitchen. He knew good things came out of there somehow, though he himself was unable to open the refrigerator and find any of them. He returned to the bedroom and, seeing Larry sitting on the armchair with one leg crossed over the other, prepared to jump. I am certain he was recalling the times Larry had him climb up his leg to get nuts. At that instant, however, Sweetie spotted a peanut on the floor, took it, and left. So much for my resolve. But the Brazil nuts I had been supplying for burying gave me an idea.

I knew that current theories of animal training were based on rewards rather than punishments. I had seen a whale in a marina in San Diego open his mouth on command to get his teeth brushed with a giant toothbrush. I had gone up to the trainer after the show and asked how he had got the whale to do that. The trainer explained how he had waited until the whale was in a desired position and, at that instant, would slip him a fish. After a number of repetitions, the whale got the idea that that position got him a fish. Then the trainer would proceed to the next stage: the desired position plus an open mouth, for example, before the fish would appear. The tooth-brushing had been the final step after weeks of patient addition of different stages to make the whole. I determined to train Sweetie to scrape his lower teeth on something that would file them down. Only after scraping would I reward him with a bit of shelled walnut.

In principle, the idea was good. Two things were wrong in practice. The first was that I did not have sole control over Sweetie's food—that is, his rewards. He was not a captive animal, and if things got frustrating enough, he could simply take his business elsewhere. True, he would have difficulty finding someone else who would crack his walnuts for him, but he might settle for peanuts if he was sufficiently annoyed at the high cost of walnuts at my place; peanuts he could handle himself. The second deterrent to the training was that I have less patience than a squirrel, and I am also a soft touch when confronted with an appeal from an animal.

The object I would have Sweetie scrape would be a Brazil nut, since a Brazil nut was something Sweetie would want to possess, though Lord knows what for, in his toothless state.

I selected the roughest Brazil nut in the package and put the training program into action. I held out the Brazil nut. Sweetie tugged and tugged; then I quickly held out a piece of walnut. He didn't want it. I had long suspected that Sweetie had a source of junk food somewhere. Perhaps it was that secretary in the Graduate School of Public Administration who had been feeding some of the squirrels Oreo cookies. I let Sweetie tug once more, and scrape in the tugging, and then gave him a Brazil nut to take away with him, a smaller, smoother one than the one reserved for scraping. He buried it under the bedroom door. He had recently taken to following me to the kitchen when I went there for nuts, and so he was near the door during his initiation into scraping. I held out the scraper Brazil again. He used leverage, pushing my hand with both front paws as he pulled with his teeth. This time, when I didn't let go, he bit me, deliberately, I think. There was more burying, trading of Brazil nuts for peanuts, juggling, and eventual departure with a Brazil. The course of training was anything but smooth.

The second day, he was evidently hungry, and the training worked as planned, with walnut bits. After looking discouraged for a while, Sweetie caught on: In order to get a piece of walnut, he would have to humor the fool by scraping a Brazil nut. Several days later, Sweetie would give the Brazil nut three rapid scrapes that would file a feather and then look up at me with an expression that clearly said, "Okay, now where's my walnut?"

I had the uncomfortable feeling that Sweetie had outwitted me. Sometimes he would balk at having to scrape the Brazil nut, but I showed strength and he would ultimately give in: *two* tiny, perfunctory scrapes and then the look. I had to give him his walnut because he had scraped the Brazil. He ate two entire walnuts in about ten installments in this manner one morning.

Things were getting more complicated with Halftail, too. She was still quite nasty. Larry and I both had wounds from her attacks, I had a bad bruise on my wrist from flinching and hitting the iron grill when Halftail attacked as I was putting out canteloupe seeds. To defend myself, I put the seeds on paper toweling and then slid the seeds off. This put a foot of paper between me and the attack. The technique worked a couple times, and Larry was emboldened enough to slide the seeds off near Halftail as she ate. Either she was surfeited or in a worse mood than usual, for she growled as the seeds came down near her and attacked the paper.

Immediately after the paper attack, Sweetie showed up and I held out his Brazil-nut scraper. He refused to scrape. I wouldn't budge. He gave his two perfunctory scrapes and I held out a piece of walnut. He refused it. After a number of repeated scrapings and refusals, I realized that he wanted a Brazil nut for burying. He evidently was not hungry after the two walnuts he'd had the previous day. Sweetie and I were beginning to understand one another. If he wasn't hungry and wanted Brazil nuts for burying, he would refuse to scrape. He recognized the rough Brazil nut I used as his scraper and associated it with bits of walnuts. If I held out a different Brazil nut, he would tug at it and scrape until I gave it to him. He fell over backward once when I released it. I would have to figure out when he was ready to eat his walnut bits. If I was wrong, out of annoyance at being made to scrape when he wasn't in the mood, or for whatever reason, he would leave me and turn to the cup, to go through the peanuts and a batch of fresh acorns Maude had brought from the Bronx. The acorns were inspected thoroughly.

Sweetie's second trapping took place in October, right after Columbus Day. It was on our first trip back to New York. We had returned not to trap Sweetie, of course, but to visit Larry's father and to pick up a car we had leased. After three weeks in Cambridge, we had to undo the results of a prolonged stay with Jeannie. She fed the squirrels at an earlier hour than we liked, since she was due at work at nine. Halftail would be around by seven, climbing up the screen to get to the glass of the window on which to tap her nails. The different hours in the different eating establishments must surely have been a source of confusion, if not frustration.

On Columbus Day, Sweetie was quite long in the tooth, and by our next trip in, three or four weeks hence, he might have a cheek infection. The hospital wasn't operating on Columbus Day, so it had to be the day after. Once again, I practiced at night to make sure both trap and I were in working order. I wanted to keep

Sweetie out in the morning until I had had time to telephone the doctor to make sure he could give Sweetie an appointment that day. I awoke early and sneaked out of bed to shower and dress, pulling the cover back up to simulate a sleeping body, in case Sweetie should peer in. But at eight-thirty, Jeannie phoned with the news that Sweetie was on the fire escape. Sweetie heard not only the ring of the phone but my voice answering. Since he knew I was up, he would paw the screen to be let in. I hung up the receiver and stood motionless. Maybe Sweetie would think he had been mistaken. Whatever he thought, he went away, to my relief. At nine o'clock, I telephoned the doctor. I could bring Sweetie in from nine to eleven-thirty, or from five-thirty to seven—not in between or there would be no doctor to attend to him.

I was ready. Sweetie evidently was not. Fifteen minutes had passed and he had not returned. I phoned Jeannie with instructions to tell her friend not to give Sweetie any Oreo cookies if he showed up at her office window. I phoned Jeannie twice more out of general nervousness, once to tell her when he finally arrived. By that time, I had made *her* nervous.

Holding a piece of Brazil nut, I put my hand through one of the doors of the trap, enticing Sweetie through the other door to get it. No problem. I let him leave with his bit of nut. When he returned for more, I trapped him. I don't know how I released the doors with my left hand without trapping my right hand, which held the Brazil nut inside. That was not the way I had practiced, but Sweetie was inside and he was less violent than the first time. I was calmer, too.

I called Jeannie. She accompanied me to the hospital, explaining that she was the aunt. Although Sweetie had no infection on his cheek, his teeth were as long as the first time and again caught on the wire of the cage. Again he uttered a little cry each time that happened, and I lifted them off the wire and pushed them back into the cage.

In the waiting room, we got acquainted with the other patients. One was a puppy with a bandaged leg strapped to a roller. Another, a wirehaired dachshund in for a shot, was held by a girl of about ten who kept kissing him. I commented to her, "He doesn't appear to lack love, does he?" She smiled. "Did you ever stop to think," I asked, "how lucky some dogs are in spending their lives in wonderful homes like yours, and how unlucky others are in not having homes at all?" The girl's eyes widened.

Just then, Dr. Rosenfeld came out to talk to me. He said he'd see what he could do. He didn't want to take Sweetie's teeth out from the roots. The roots were deep in the jaw, making an operation dangerous; and Sweetie would be left maimed. I suggested that less anesthetic be used than last time, as it had taken Sweetie so long to recover from the effects. I was to call for him at five-thirty.

At five-thirty, Sweetie was stirring in the cage. Dr. Rosenfeld said that all he could do was file his teeth, and file them he did —almost to the bone. Last time he had left about an eighth of an inch. "If we're going to do this regularly," Dr. Rosenfeld said, "we should have a sliding scale. What did I charge last time?" "Twenty," I said. "Then let's make it fifteen." My initial shock that a member of the medical profession, albeit the animal branch, had a sliding scale was tempered by my realization that, in fact, less service had been rendered: This time Sweetie had not been given an antibiotic, since he had no infection.

At home, I put the cage on the desk. Sweetie regarded me quietly. Then he made as if to get out. I lifted up a door. He seemed alert. He jumped to the windowsill. When he found the exit blocked, he climbed up the iron grill, climbed back down and jumped to the floor, jumped back to the windowsill, then headed for the living room, possibly to seek some other way out. Larry wanted me to make certain he was in full command of his faculties before I let him out of the apartment into the world. "You should have released him in the garden," Larry said. He was worried that Sweetie might fall off the fire escape on the way down. It was too late for the garden; Sweetie was running around the living room. Anyhow, I think these creatures are able to climb down fire escapes blindfolded. I opened the window.

The next day started off wonderfully. Halftail arrived at eight-thirty and Sweetie arrived a few minutes later. He picked up some acorns and peanuts from the sill, and even tried to scrape one. Then he jumped onto the desk and waited for me to wait on him. He grabbed a walnut from my hand and tried to scrape it, but soon dropped it. I cracked the walnut and handed him half in the shell. He had difficulty getting the meat out, so I broke the rest into pieces. Sweetie carried out four walnuts in halves and pieces. From the short intervals between his appearances, I suspected that he wasn't able to manage the halves, and that he might be storing even the pieces somewhere for times when I might not be available for service, because he realized he couldn't manage anything else. Four walnuts was double his normal capacity. I had told Jeannie that I wanted to give Sweetie one good meal, to regain his trust before I left for Cambridge the next day. Then I wanted him to fend for himself on peanuts, acorns, and almonds. The small amount of grinding these nuts required might lengthen the intervals between trappings. But seeing how much difficulty he was having, I left Jeannie a note asking her to help out a while longer. A man I knew years ago, a kind of mentor, had once told me that, of his three sons, he loved the youngest most. The youngest had been a very sickly child. "One always loves most the one who gives the most trouble," he said.

TRAPPED AGAIN

Back in Cambridge, we were struck by the good life the Boston-area squirrels had: millions of acorns—big ones—and few cars. The squirrels in our garden had time to play. They tore back and forth across the telephone cable, acrobatic marvels. The ones near Larry's Astrophysics Center had evidently put in a great deal of time burying—there wasn't a decent acorn left on the ground; only cracked ones remained. One block from our house, though, there were so many acorns I began to gather some to bring back to New York. I must have collected fifty pounds.

Despite the easier life in Cambridge—not only for squirrels but for people, too—the squirrels were more skittish. It is possible that because they had less need of human beings they avoided contact with them. I could not succeed in getting a single Cambridge squirrel to come up to me, no matter what I offered. (A few years later, I would find this to be true of Princeton, too, where again there was a plenitude of acorns.) They were quite unlike New York squirrels, who come right over to perfect strangers and strike up an acquaintance.

We were in New York twice before Christmas. Each time, Sweetie would come right up to the apartment, exhibiting no resentment that I could detect. I had put him on almonds, most of which he was able to shell himself. He took filberts, but they were clearly for burying only. He took acorns, too—or maybe it was Halftail who took them. Despite dental problems, Sweetie was still tough and aggressive toward Halftail, chasing her every time he caught her in the apartment. Sweetie was beginning to look like a tank, heavy and solid the way Genius had looked in his prime. A false front. It was his winter fur. There were several new squirrels, a small gray, a charcoal-colored one, and a black with one ear pressed against his or her head. The newcomers were all shy and appeared to be young.

Sweetie's third trapping took place Christmas week. We had arrived in New York on the twenty-second, and had been feeding Sweetie for a week. While I was cracking a walnut for him, he was trying to tear it out of my hand. He had no patience, and almost lost his nose in the nutcracker. Larry remarked that perhaps all animals that are prey are impatient. The cats he had had all through his childhood, on the other hand, in stalking birds, would remain motionless with near-infinite patience.

On the morning of the twenty-eighth, I was mentally and emotionally prepared for trapping. I had practiced the night before and had made an appointment with Dr. Rosenfeld. It was a cold day, however, and Sweetie was not likely to venture out early. We wanted to have Sweetie attended to that day, since Dr. Rosenfeld would be off the rest of the week. If not, we would have to stay on in New York longer than we had intended.

At quarter past eleven, with no sign of Sweetie, I canceled his appointment. Dr. Rosenfeld would be leaving the hospital by about noon.

Several minutes before noon, Sweetie arrived. I gave him a quarter of a walnut on a shell and he left. I hastily telephoned Dr. Rosenfeld and asked him to please wait until noon. Sweetie returned. Everything was set up. Walnuts were in the trap. But Sweetie was wise. He wouldn't come in. He kept wandering around outside, picking up bits of acorn on the outer sill, popping in the entrance but popping right out again. I coaxed him in with a bit of walnut, which he ate on the desk, but he would not go near the trap. When he left with an almond, I thought all was over. But he returned and went through the same shenanigans. Unable to stand it any longer, I reached over to the sill and grabbed him around the waist. He grabbed the burglar grill. I couldn't pull him off; he hung on with what must have been more than twenty pounds of force. I pried his fingers loose from the grill. His heart pounded against my hand. I stuffed him into the trap.

Larry and I both shut the doors. A bit of fur from Sweetie's tail remained outside. Larry phoned Dr. Rosenfeld while I charged down the stairs with Sweetie.

Dr. Rosenfeld was outside the hospital in his street clothes, throwing out the garbage. Sweetie was taken into the operating room immediately. I left three walnuts for him and went away, totally depressed. I could still feel that warm little body, the pounding heart, the fright I knew he must have felt to enable him to cling with such strength. I felt rotten.

I called for Sweetie at four o'clock, as had been arranged. I paid the fifteen-dollar fee—there was no mention of a sliding scale—wrapped the cage in newspaper, and carried it out. Sweetie stirred as I walked, then fell asleep again.

In the house, I opened the doors of the cage so that Sweetie could emerge. He was spaced out. His legs were spread wide, he couldn't jump, he had trouble climbing—he kept falling. I feared I might have injured him when I grabbed him. He climbed up the rubber plant, and slipped. I caught him and set him down on the floor. Again he tried to climb up the plant and again I caught him as he fell. Then he tried to squeeze under things: the radiator, the plant stand. He almost lifted the rubber plant in his effort to get under its rolling stand, a plant that weighs perhaps thirty pounds. Then he took to running backward. He seemed to be having nightmares. His running backward resembled movements he had made in resisting being put into the trap. It was as if he were reliving the episode.

Meanwhile, it was getting darker outside. By five-thirty, it was completely dark, and Sweetie was still pushing his way into corners backward and falling asleep. He crept behind the bookshelves near the radiator, then between the file cabinet and the bookshelves—dark places. I put soft rags into the trap, gently lifted the sleeping Sweetie and laid him on top of the rags, carried the trap into the living room, and set it down on the floor near the coffee table. I blocked off the streetlights with a large floor pillow, turned off the radiator, and canceled our dinner appointment with Joan and Dick Silverman. We donned coats and ate leftovers in the freezing kitchen. We kept Sweetie's part of the apartment dark and quiet. Lord knows whether that was what he was used to at home, but it seemed appropriate.

I checked on Sweetie periodically, getting down on my knees to listen to his breathing. On one such check, I had difficulty hearing Sweetie. Something was interfering—a loud shushing that sounded like a train. I couldn't imagine what machine might be making that sound. The radiator was off. Then I realized—it was Sweetie! He had the heavy breathing of a masher!

I checked several times during the night. When, toward morning, I heard Sweetie sniffling, I turned the heat on.

Larry had set the alarm for six-thirty A.M. Thinking it would still be dark then, I changed the setting to seven. It was dark at seven and there was no sound from the living room. At quarter past seven, I investigated. I feared Sweetie was dead—smothered by his rag mattress or O.D.'d from anesthetic. I got down on my knees and peered into the cage. Sweetie was asleep.

I carried the cage into the bedroom, placed it on the desk, and opened one end. I also opened the window and cleared the exit. Sweetie awoke. He tried to get out the closed end of the cage. He soon found the open end, however, brushed past my proffered walnut, and took a giant leap to the railing of the fire escape. I gasped. He stood on the triangular corner observation spot for perhaps a minute, surveying the scene below, then ran along the railing and disappeared. He did not return later when the others came. Our relationship seemed at an end. I was depressed.

But Sweetie did come back. The next morning, he knocked on Jeannie's window. She gave him half a walnut and "sent him upstairs." Larry thought he had seen Sweetie a few minutes ear-

lier; so did I. As there had been only acorns and almonds on the
sill outside, he may have tried to do better down at Jeannie's.
When he came upstairs again, he put only his head and shoulders
in the entrance to take a piece of walnut from my hand; he would
not come in. He did not hate *me* at least, only what happened to
him inside my house.

The third morning after Sweetie's night at our house, he again
hovered in the entrance. But when he saw me approaching with
a walnut, his eagerness propelled him inside and all over my hands.
Thank goodness he didn't associate my hands with his awful
experience.

Gradually, over the next week, affairs returned to normal. On
the last day before our return to Cambridge, Sweetie arrived early
and was on the desk insisting on walnuts. He would take nothing
else. He waited, clicking with annoyance at the delay, while I got
walnuts from the kitchen for him.

In Cambridge, I had time to think. Apart from any effect the
trapping might have on Sweetie, it was having a terrible effect on
me. I was haunted by that pounding heart in my hand. Even
though Sweetie had evidently forgiven me, I couldn't see going
through it all again every couple of months. And it would surely
get worse as Sweetie learned each new ruse I would be forced to
employ. Larry suggested I contact B.F. Skinner. "If anyone knows
how to train animals, he does," Larry said, regaling me for the
thirtieth time with the miraculous feats he had seen pigeons
perform under the tutelage of trainers schooled in Skinner's tech-
niques. Starting from scratch, in minutes pigeons would learn to
peck at a red square or a green circle, reinforced by a pittance of
a pellet. Skinner was at Harvard. We were at Harvard. Why not
consult him?

I telephoned the psychology department. The male secretary
informed me that as Skinner was now professor emeritus he did

not come to his office regularly. If I kept telephoning the number he would give me, I might reach Skinner's secretary, who would set up an appointment. I would have preferred to simply drop in at Skinner's office—in William James Hall—to make an abrupt brush-off more difficult, but Skinner's irregular hours appeared to rule that out.

The first time I called the number, I was prepared to make an evasive appointment speech to a secretary. An elderly, nonsecretarial male voice said, "Hello?" "Am I speaking to Professor Skinner?" I asked. "You are." I had little choice but to plunge right in with my problem, unrehearsed. Skinner not only didn't find me or my problem peculiar, but responded as if he encountered such situations every Monday and Thursday and gave them serious consideration. He was most friendly.

I had made several mistakes. When Sweetie was scraping against the Brazil nut, I should have been rewarding him with more walnut for more scraping. A Brazil nut wasn't much good in any case, according to Skinner. A piece of carborundum was what I needed. Skinner volunteered to design a mechanism that would release a piece of walnut to Sweetie when he scraped on the carborundum.

Skinner had a second suggestion, however: that I have Sweetie put to sleep. He might die a humane death then, rather than the painful one that would surely be his if my efforts were unsuccessful or if he stopped coming to the apartment. Taken aback, I mumbled something about not wanting to play God. Skinner pointed to the obvious fact that I was already playing God. I thanked him, said I would think about both his suggestions, and hung up. A few hours later, however, I was thinking only about how I could get the mechanism made if Skinner should design it. Would Nick Rinaldis, the machinist at Rutgers, be able to do it? I might have to trap Sweetie a couple of times before the contraption could be completed.

The real God decided Sweetie's fate in celestial fashion. On January 20, twenty-seven inches of snow fell on Boston. New York had almost that much. Jeannie reported to me later that Sweetie and Halftail did not appear for five days. Then, on February 6, there was the great blizzard—the one that was compared to the blizzard of '88—with its record snowfall.

When Larry and I went back to New York on February 18, Jeannie reported that she had not seen Sweetie or Halftail since the blizzard. It was not a question of snow on the route, we thought, since some of the regulars were at the window, the three that had joined the team since we left for Cambridge in September, the black one with an ear flattened against his head, the charcoal one, and the young gray with Halftail's temperament (he growled when a nut was put down near him while he ate). Jeannie hadn't put out nuts for them and they had come upstairs. We concluded that Sweetie and Halftail were snowed in. I had long suspected that they were living together, as they usually arrived at the same time. If the entrance to their abode was blocked by snow—if they lived in the trunk of a tree, for example—that would explain their five-day absence after the first big snowfall, and their longer absence now. My guess was that there would be plenty of nuts in their nest, but they would be whole filberts. Halftail would be able to survive being snowed in, but Sweetie? I couldn't see Halftail shelling nuts for him. She might shell involuntarily, if Sweetie grabbed shelled nuts from her.

By March 5, a month after the blizzard, we had gone back to Cambridge and returned to New York once more and Sweetie had not been seen. The other regulars were on the fire escape. The charcoal one looked terrible; he had a kind of mange. The little gray meanie was eating only peanuts; hard-shelled nuts were apparently beyond his abilities. Halftail was there. She had appeared at Jeannie's window three weeks after the storm, looking none the worse for her ordeal. It was now a week since she had resumed her visits. Still no Sweetie.

I stood looking out the window, hoping Sweetie might be alive, or that he had died a painless death, going numb and freezing to death in a short time. I knew that was improbable, though, and most likely he had starved to death. But at least he didn't have the pain of his tooth penetrating his brain case.

As I stood thinking about Sweetie, I saw a gray squirrel approaching, a large one, larger than the little mean one, and not in very good shape. I prayed it might be Sweetie. I'd feed him a dozen walnuts! It wasn't Sweetie—the ears were notched. Those notches looked familiar—a large, smooth one on the left ear and a tiny one on the right ear. Then I saw what looked like a tiny sty in the bottom rim of the right eye. And the front left paw! Philip! Clubfooted Philip! The Lord taketh away and the Lord giveth. I hadn't seen Philip in a year!

I opened six walnuts for him. After Sweetie, Philip was the next best sight I could have seen. No, not quite—Genius would have been next. But there was plenty of joy in seeing Philip. For his part, he acted as if he'd never been away. He sat in his accustomed spot right in the exit, hunched under the window, and ate five and one-half of the six walnuts. He took nothing with him when he left, not the remaining half-walnut or a filbert for the road.

I had no idea how Philip had been managing all this time. It had been months since Jeannie's friend Rosalie, who used to feed Philip Oreo cookies at 5 Washington Square North, had taken a job in Boston. Her replacement hated squirrels.

Philip showed up again two days later. He was able to crack filberts unassisted. After a number of them, however, when I handed him one, he dropped it. He wanted me to crack it—which I did. In the interest of scientific honesty, I must admit that this may not have indicated his remembrance of things past, as I had opened a walnut for him just prior to his request for filbert cracking. Again he left emptymouthed. I didn't see him the next time we were in from Cambridge, at the beginning of April, or the following time, at the end of April, although Jeannie said he had been around. I have not seen him to this writing. It was as if he had come just to comfort me in my loss.

17. A HARD LIFE

SQUIRRELS, LIKE PEOPLE and dogs, lead different lives in the suburbs. And people's lives are different in ways other than the obvious. During our sabbatical year at Harvard, we lived in a lovely neighborhood in Cambridge, near Fresh Pond. On our street, Larchwood Drive, there were only one-family houses—some were mansions—spaced far apart, with luxurious gardens. In the fall, our garden was covered with leaves, and neighborhood boys rang our bell to ask if they could rake the lawn. In winter, after the first snowfall, boys came asking if they could shovel the snow from our driveway. They raked and shoveled for pocket money. Boston's inner-city boys, who undoubtedly needed the money more than these neighborhood boys, didn't know of the existence of these lawns and driveways that could be raked and shoveled, or, if they knew of them, had neither the rake and shovel nor the carfare to get to them. I was delighted, therefore, after the second snowfall, to have my doorbell rung by two boys with accents that stamped them as nonlocal—with a *snow blower*. A third member of the team stood apart from them, on the street. An adult, he was possibly the source of the large capital investment. I hired them. It took them fifteen minutes to do our driveway and front walk.

I estimated that they could make about sixty dollars in an hour. That, and the thought of Larry's bad back, overcame our guilt at not doing the shoveling ourselves. I actually enjoyed shoveling snow. I needed the exercise, and having lived in apartments all my life, I liked having my own snow.

Suburban dogs live a suburban dog's life. In the city, dogs are alone a great deal. To go out, they must await their owner's return from work, and then "out" is usually on a leash. Depending upon the owner and the neighborhood, the amount of social intercourse with other dogs varies, but it is ordinarily restricted. The dogs on Larchwood Drive, on the other hand, would take walks and pay social calls all day long. Jason, a golden retriever, would visit me every time he saw me outside. After a few vigorous rubs from me on his golden fleece, he'd take off for another call. A Malamute named Bandit would stroll over to Star Market each day and come back with a bone. Very few of the dogs were kept indoors. I saw none tied in their yards. I rarely saw any on leashes. Fresh Pond was jogging territory, and it seemed as though half the joggers were dogs.

The dogs that went to Harvard had an especially fine time, I thought, a kind of student life, congregating with cronies in Harvard Yard while their owners were in class, making friends with people like me who would throw sticks for them or play tug of war.

For squirrels, life in the suburbs is probably ideal, or so it seemed to me. They have all the advantages of life in the woods, large trees, all the acorns they can eat, and no predators. Cars were few, and those few moved slowly, and even the Larchwood Drive dogs weren't interested in squirrels.

The good life showed in the squirrels' behavior. They would have nothing to do with us. They didn't need us, as our Washington Square bunch did, for daily handouts. (In fact, thinking about our relations with our New York squirrels, I could sympathize with the heiress who always suspects people are interested in her only for her money.) Larchwood Drive squirrels had time to chase birds. At least, that is what I concluded they were doing when they raced through completely bare trees, not a sprout of anything edible on them but with a bird in the direction of the sprint. The bird (often a blue jay) flew off with time to spare and not much flurry, as if participating in a game. In New York, they had jealously dive-bombed our panhandling squirrels.

People act differently toward squirrels, too. This came as a surprise to me, even though I had known about it intellectually, through having had to borrow the squirrel trap. In the city, squirrels are friends, entertainers, charming clowns, gymnasts, nature's creatures to be enjoyed and pointed out to small children, who are given peanuts with which to feed them. In the suburbs squirrels are the enemy. They must be kept out of attics, trapped if they get in. They must be prevented from getting the food that is put out for birds. No food is put out for *them*. No wonder they would have nothing to do with us.

I was torn between my newfound suburban mentality and my native city mind when I saw three squirrels outside the kitchen window, eating the seed and sunflower mix I had bought at Star Market and thrown out on the snow for the birds. Mrs. Kepes, whose house we had subleased, had gently warned me not to encourage the squirrels. She, too, fed the birds and worried about her attic. I probably would, too, if I had an attic.

Larry thought the Larchwood Drive dogs looked bored. He thought the best animal life was had by Kalmbach, our cat neighbor in Aspen, who hunted each night in the warm months—he'd bring us presents in the morning, dead birds or field mice—and moved in with a family he'd adopted when winter came and the hunting got harder. (I had named him Kalmbach after the Kalmbach of Watergate fame, who knew someone was telling the truth because he "looked into his eyes." Kalmbach the cat would jump onto my lap, put his paws on my shoulders and his face right in front of mine, preventing me from reading or watching television or doing anything except look into his eyes.) The animals who aren't free like Kalmbach also have a great life in Aspen. Dogs visit and socialize as in Cambridge, but on the way home they'll take a swim in a pond or a stream. And on weekends, they hike in the mountains. For the expression of joy in being alive, there is nothing to match a dog on a hike. I've had the great good fortune, several summers in a row, of having golden retrievers—three—as neighbors. Their owners were only too happy to let them accompany us when Larry and I went hiking. I announced to Larry recently that if anything were to happen to him, I would marry a golden retriever.

I have often brooded over the way the luck of the draw operates so much more relentlessly in the lives of animals than in the lives of humans from almost the moment they enter the world. The home in which a dog finds himself is determined purely by chance; the dog doesn't have the genetic tie most humans have to a particular home, the home of parents. And then the dog lacks even the meager means a human has of altering a desperate plight —the appeal for help. The embodiment of these thoughts, for me, was a dog named Tinker. She had been snatched from the jaws of death in Hong Kong by the young couple, Sara and Tim, whose Hindu wedding I had attended years earlier. When I met Tinker that sabbatical year at Harvard, Larry and I, Sara and Tim all looked at her and wondered, "What is the probability of a five-week-old puppy in Hong Kong, in dog-eating season, ending up in Newton, Massachusetts?"

18. MOURNING BECOMES HALFTAIL

FROM AS FAR BACK as I can remember, I had considered myself to be my father's daughter. Knowing he had wanted a son and had been disappointed twice, I tried to make it up to him, to be the son he wanted. I was thrilled when he made wooden guns for me from orange crates; they were slingshots, essentially, which used cardboard squares as ammunition. I took apart flashlights and clocks, things I thought *he* was interested in. I believe now that I went into physics to show my father I could do what a boy could do. To attribute my attachment to my father solely to some sort of Electra Complex, however, is to oversimplify. I had a good deal more in common with my father than with my mother. He was an intelligent man, interested in intellectual matters. He had been educated by his father, an intellectual, in Russia, until the age of sixteen. Then he was sent to America—there was only enough money to send the eldest son to a university and on to medical school. My father was sent to join another brother, my uncle Abram, in New York. Uncle Abram proved not to be the greatest influence my father might have had. My father was not encouraged to go to night school, as some young immigrants were. He wound up in the garment industry, and remained there until he

151

married, at twenty-three. Then he opened a succession of small grocery stores—a succession because he had no success with any of them. He was antidiplomatic. If a customer informed him that she could purchase an item at a lower price elsewhere, my father would tell her to go elsewhere then. The best time in his working life was during World War II, when he was a welder on aircraft carriers at the Brooklyn Navy Yard. He was happiest and made his highest salary then. After the war, he returned to the garment industry; he had had enough of the 361 day work year, 103 hour work week, which characterized the family grocery business at that time. He never made more than a very modest living at any time, and was considered a failure by the family, by my mother, by himself, by everyone but me. I sensed his frustration and his misery long before I was able to understand them, and tried to make up to him what life had cheated him of. Any time I achieved some success in school, my father was the one with whom I would share it.

My mother was, and is, a simple woman. Intellectual snobbishness on my part led to my valuing little other than intellect in my youth, and to finding my mother weak and often whining. I had known from early childhood that my parents were terribly mismatched. What I hadn't known is that one person can squelch some aspects of another's character, leaving a lopsided or partial person. When my father died, my mother was transformed. She stopped whining. She joined a Y and a senior citizen's club and began the active social life she had always yearned for but had been kept from by my father, by his contempt—which he made no attempt to hide—for my mother, the friends she tried to make, and all the relatives on my mother's side of the family.

I didn't mourn my father. He had died, at seventy-three, in perfect health. He was extremely strong, and the fastest walker I have ever known. He had developed a heart condition a few years earlier, a biological fact that he refused to acknowledge. He died ignoring it. I was surprised that I didn't grieve. I accepted his death as the end of his life, and accepted the hour that that life had ended. Feeling uneasy about my acceptance, after observing friends who did not accept their fathers' death with equanimity, I concluded, with a great deal of rationalization, no doubt, that deep grief comes not necessarily from deep love; there may be other sources, such as guilt or dependence. I had neither of those. I felt no guilt; I had been a good daughter. And I felt no dependence. I certainly loved my father. The "true" reason for my not grieving probably lies in my personality; I accept certain things— although what I accept and what I do not, I can't say, except possibly that those I accept are things over which I feel I have no control, and could in no way alter. After my father's death, I became friends with my mother.

A change came over Halftail after Sweetie's death. She mellowed. She became much gentler. She stopped grabbing. I was able to give her nuts by hand. She still ate everything she was given —acorns, seeds, whatever there was—but she ate them more slowly. Halfy got special treatment from me, walnuts when others got plain fare, and a welcome inside whenever I saw her on the fire escape—no office hours for Halfy. She came at all hours, but when she was really hungry, she came at opening time in the morning. I liked the way she checked, when she came at off-hours, to see if she could get in. Usually, the filter was outside the window when the window was closed—for protection. When Halfy appeared on the outside sill and faced the filter, she could not be certain what lay on the other side: the glass of the closed window, the screen, or an open entrance. She would climb onto the filter at the position of the entrance and reach out a paw to touch glass, or wire mesh, or nothing; if nothing, she would climb over the filter and into the house. Halfy appeared to know that she was getting special treatment, always seeming unperturbed by the tumult of the other squirrels fighting over the food, above the fray, confident that she would get hers. And she looked marvelous—the healthiest squirrel in New York.

She did exhibit remnants of her old character on rare occasions. Once, while she was sitting munching in her usual position on the extreme left of the fire-escape railing, I started to clean up by taking in shells that lay on the outer sill. Halfy pounced from the railing right next to my hand, a shadow of her former attacks. I never did figure out why some squirrels—only a few—growled and lunged when I put food near them as they ate. I could understand if I had been taking food away.

A couple of months after Sweetie's death, Halfy took up with a black squirrel. In addition to Halfy, a little gray was coming regularly, as were two blacks, a male and a female. Halfy and the black male would sometimes show up for breakfast together. While downstairs in the garden, on my way to the laundry room one day, I saw the two of them racing through the trees. I don't remember who was in front.

Nine months after Sweetie's death, Halfy was so transformed that she sat on my lap eating half a Brazil nut. It is true that she wanted to be near the other half, which was in my hand, and that she soon became uncomfortable facing me and turned away after a few mouthfuls, but sitting on my lap facing in *any* direction constituted an enormous change.

Halfy's lover, whom I named Blackbuster, seemed a light-hearted fellow. I saw him romping in the garden with a Kleenex, jumping backward and tossing it up, ultimately shredding it. He seemed especially flippant to me at that moment because I was engaged in sweeping up nutshells—many of them his—an activity I had voluntarily taken on, on laundry days, with equipment borrowed from Superintendent Lassiter's office, as a conscience-easing use of the time before the laundry could be transferred from washer to dryer, and one which would forestall Mr. Lassiter's complaints about my feeding the squirrels. Returning the broom and dustpan, I chatted for a while with Mr. Lassiter's assistants, all of whom liked animals. I learned that Lester, the night porter, regularly dug up the nuts the squirrels had buried, so that he could personally hand them out again for reburial.

Blackbuster came to the apartment approximately every third day. He had a decided preference for filberts, and accepted nothing less. He was extremely gentle with us; a toucher, he put a paw on my hand when he took a nut. I liked Blackbuster. Apparently, Halfy didn't, though, for the affair didn't last. I saw him on the fire escape, trying to force his attentions on her. He didn't know that one does not force attentions on *Halfy*. She growled him away, jumped over him to get into the apartment, defying Newton in the act by doing a half-gainer in the air, which, I am convinced, did not conserve angular momentum. Blackbuster continued to come to the apartment, but not with Halfy.

Meanwhile, Halfy had learned how to pull the filter out of the window. The first time she did it, it was justifiable. We were sleeping late on a Sunday in March and ignored a banging at the window at eight-thirty. At nine, she ripped the filter out. I got out of bed and went to the window. Halfy stood looking at me expectantly. It was snowing lightly. She had never pulled out the filter before. She tugged and tugged until it simply came free of the wooden slats on the window frame. She must have been desperate. Did she remember the previous year's snowstorm and the end of Sweetie? She ate as though she did. She pulled the filter out twice more in the ensuing weeks, each time while I was in the house, making me conclude that she did it when she knew I was inside —having seen me, perhaps—ignoring her.

I kept altering my opinion on whether Halfy, Genius's daughter, had inherited Genius's brains. On one occasion, I thought she was positively Talmudic. About half a dozen squirrels had been coming regularly. In general, the males preferred filberts while the females ate everything given them. To minimize fighting, I had been putting assorted nuts outside for the females to get at easily, reserving the cup for filberts, for which the males could come inside. Halfy developed her own pattern of dining. When she dropped in—literally—she would bypass the nuts outside, make no attempt to take any from the cup, but would drop down onto the chicken wire over the soil of the rubber plant and wait for me to hand her nuts there. I followed her lead and began to leave nuts on the rubber-plant wire for her to eat inside, under the protection of the law, so to speak. Sometimes, she would transport one of these nuts to a point under the desk and eat it there. When a male entered the room, she would stop and get into a ready-for-flight position, which she maintained all the time he was getting his rations, relaxing into "at ease" and continuing her meal only when his tail disappeared out the exit. This time, Halfy apparently required a filbert for the road, and there weren't any on the plant. I spied her hanging upside down from her back feet from the burglar grill, taking one from the cup with her teeth. It was as if she believed she could take from the males' cup provided she did not touch it with unclean female hands. It reminded me of the story about an injunction against raising pigs on the soil of Israel, circumvented by some enterprising souls who built wooden platforms on which to raise their pigs so that the pigs could not be said to be on the soil.

Halfy's braininess at the cup was canceled out by some denseness at the door. One morning, she appeared outside when the screen was closed and the filter happened to be on its side, inside the screen, directly in front of the entrance. When I opened the screen, I didn't move the filter but simply opened the screen wider than usual, enabling Halfy to come in through an opening to the left of her usual entrance. She took the nut I handed her, then stood calmly in front of the filter, waiting for me to move it so she could get out. She had no recollection of the opening to the left, through which she had entered seconds before. Strange. Others had exhibited this same denseness in not remembering where they had entered. Yet I had seen evidence that memory, rather than sight, was sometimes employed. (One individual crashed into the burglar grill on leaving; I had opened it for repair work after he entered, compressing the bars in front of his normally open exit. Another inserted a snout into a cup that was no longer there.) I wondered whether there might be a spatial problem—the new opening was on Halfy's right when she entered, on her left when she exited—or whether the location of the exit was so firmly imprinted by so many exitings that a change could not compete.

The next morning, Halfy pulled the screen out shortly before opening time. She jumped down onto the rubber plant and noisily shelled an almond I had left there. I pretended to be asleep. When she finished the almond, she jumped down onto the floor and cautiously explored the room, her tail flicking up and down. She examined my shoes, *The New York Times*, the bottom of Larry's shirt, which was hanging from the closet doorknob. Finally, I rose, feigned surprise at seeing her next to the bed, and got her her breakfast.

MY SON-IN-LAW THE SQUIRREL

I was beginning to fear Halfy would never get married. Despite improvement, she might simply be too unpleasant to maintain a lasting relationship. But in the middle of May, Halfy brought home a boy. He was an exceedingly handsome fellow. I thought so before I knew he was her intended. He was big and sturdy, very virile-looking—if such a word can be applied to a squirrel—with no signs of imperfection, a magnificent tail, and an interesting face. His nostrils flared, like a horse's. The handsomest squirrel I had ever seen. He was also quite accomplished. Brazil nuts and walnuts lay in halves on the sill. The handsome fellow took the first half he came upon, a Brazil. In taking it, however, he spotted a half-walnut. Highly excited, the Brazil nut in his mouth, he hooked the edge of the walnut shell with an available tooth that was in front of the Brazil, and left with both nuts. He was the first squirrel I had seen manage two nuts—though, in fairness to the others, it must be noted that these two were not whole nuts, which made the feat considerably easier.

That a courtship was being conducted was apparent: There was no clicking on his part when Halfy took nuts from the cup; she ate, calmly, in his presence; he walked after her, slowly, while she went about her business of getting and eating nuts. I felt like a prospective mother-in-law, approving of the groom-to-be.

During the next few days, I plied Husband, as I started to call him, with walnuts, to keep him coming around. (I was prepared to furnish a dowry of one thousand walnuts.) He took them from my hand gently, after a little hesitation.

When the honeymoon was over, Halfy had to eat on the fly, as she had in Sweetie's time. She regained her title of Fastest Jaw in the East. Employing an adaptation of Genius's old Dodg'em strategy, she would lure Husband to make threatening gestures on the left side of the windowsill by a slow approach on the left; then, with the speed of light, she'd get her nut from the right. I provided an assist by waiting near the window and thrusting a half-walnut out at her as she flashed by. I made sure Husband got one immediately afterward, to avoid domestic friction. Half a walnut and he went sniffing after Halfy's behind. He seemed a sexy squirrel. He *looked* sexy.

At the end of June, shortly before we were to leave for Aspen again, I became convinced that Halfy was pregnant. She seemed slower and heavier and more cautious—she crept down from the fire-escape railing along the vertical bars, whereas she had formerly jumped from the railing to the windowsill. It was about time, too.

Husband was showing up occasionally with one of the other ladies.

Jeannie was given special instructions regarding Halfy, because of her condition. Halfy was to get walnuts; they were to be put out toward evening, as Halfy tended to come around then. Jeannie was quite skeptical about Halfy's character change and still called her Crackpot, the name she had given Halfy when Halfy had attacked her a few summers earlier.

Jeannie wrote to us while we were away, but sent no birth notice. We returned after a month, and when Halfy didn't appear our first day back, I had visions of her at home in the nest tending her young. She appeared the following day, however, and she wasn't pregnant. Nor was there any sign of enlarged nipples from suckling. She jumped down onto the rubber plant and waited for me to hand her a nut.

Throughout the rest of the summer, I would see Husband occasionally. Halfy herself came regularly, but only for one or two filberts, which she was undoubtedly burying. She was merely maintaining contact, and getting her meals elsewhere—from the Danish woman in the Mews who supplied only walnuts, I suspected. Finally, Husband stopped coming. I don't know what happened. Nor did I ever learn anything more about the false pregnancy. Halfy was pushing five and had never been a mother. She'd probably have made a rotten mother anyway.

Halfy became my favorite. Compared to the way she was when we first met her, she was a new squirrel. She rarely grabbed. On the few occasions when she did (once, she thought I was taking a nut away) it was without a growl. She took nuts from our hands gently. She never went so far as to become a toucher, but she did permit me to touch her. That is, she didn't sock me when I ran my finger down the side of her back. And she was extremely considerate about waking us in the morning. If we weren't up when she arrived, she either waited quietly on the fire-escape railing or went away and returned later. Only when she saw or heard us moving about did she stomp on the aluminum plate on the window frame (installed to protect the wood when new windows were put in in the bedroom). Then I had to hurry and open the filter or screen, talking to her while I put on my slippers because she got more excited by the second. I sometimes feared she would burst before I could get to the window. She helped me, of course, by tugging at the screen with me. In short, she acted just the way her father, Genius, used to. Although none of the squirrels had patience, Halfy and Genius probably had the least.

The last time I saw Halfy was just before the summer of 1981. Somehow, several weeks before her last visit, I knew our friendship was winding to a close. I knew it because she was acting the way Genius did before he disappeared. She didn't come every day, only about every third day, and then in the afternoon, when no other squirrels were around. True, she had been doing that for some time—and I had been giving her a rousing welcome no matter *when* she showed up—but her manner had changed. Where before it had seemed to me that the special treatment she got was noted and gave her assurance— and may have contributed to her mellowing—toward the end she seemed over-gentle, even sad. How could I tell? I can't, of course, adduce scientific reasons for my belief. But can't one tell when one's dog is sad, guilty, or even embarrassed? I don't claim to see downturned corners of lips or lack of sparkle in the eyes or slackening of cheek muscles in squirrels, and so I am hard put to describe sadness in Halfy. I simply thought she looked sad. And the reasons for her sadness, I thought, were a new black female with an ear like hers, and the little adolescent male, Boldy. Halfy was still dominant over them; they ran from her. But could she remain dominant? And was it worth the hassle?

* * *

Both Halfy and Genius had seemed to me to be depressed toward the end. They may not have been depressed at all. Depressed was my interpretation of a manner that may simply have denoted old age in a city squirrel. Halfy and Genius were both about six, by my reckoning, and although the pet squirrel I had read about lived to the age of twenty, the one I *knew* about died in her bed at six. But if I pursue my conviction that Halfy and Genius were depressed, how do I account for the depression? I fear I can do that only by exercising a great deal of anthropomorphism. Genius, I knew, had suffered a loss of status. Halfy, it seemed to me, was fighting to maintain hers. I link depression and status for squirrels because that is what I do for humans.

19. BLACKBUSTER

AT ONE TIME, I had feared black squirrels would take over Washington Square Park. Seeing their number grow so rapidly from that single one I'd seen in the 1960's, I thought they must be more aggressive, or stronger, than the grays. Our squirrels lived on the other side of the park, apparently, since the vast majority were gray. I suppose we had about half a dozen black visitors in all, over the years. Blackbuster was a toucher. As was the case with a few others, however, he was gentle with us and tough with his colleagues. When he was around, the ladies had to hop to it. He learned to tear the screen out of its moorings when we ignored his signals to enter; he would drop down onto the floor and stare at Larry or me until we got him a filbert. He ate only filberts.

One day, he went off filberts. He wanted something else. Walnuts. He ate only walnuts for a few days. Then he didn't want walnuts, either. We couldn't figure out *what* he wanted. I offered a peanut; he slapped it out of my hand. He left, having taken nothing. A few days later, he showed up in terrible shape. He had a pink pimple under his left eye; it looked like an infection from a wound. He may have had mange, or lost a lot of winter fur, for pink skin showed through, making him look mottled. It was about the right time of year for losing fur. This time he stayed around for two or three days, complained about the food, then disappeared for a week. When he reappeared, he looked worse than before. The pimple was gone; in its place was an open sore. He was limping, and he couldn't eat. He refused filberts. He tried to suck on pieces of walnut, but was unable to use his teeth—if he had teeth. I offered some food I thought he might be able to eat, a dried cherry and a piece of a ladyfinger. He sniffed at them, discouraged, and left.

It must have been hard for a limping Blackbuster to reach our apartment. When the ivy on our walls was cut down—for fear it was providing a ladder into upper-story apartments for humans as well as for squirrels—the squirrels adapted by climbing up the outer grill on the ground-floor window and leaping onto the bottom rung of the fire-escape ladder—more than six feet away—by ricocheting off the wall about two feet closer to the ladder. It seemed an impressive feat, this leap so many times their own length; in fact, their length is not a significant factor: Almost all animals jump roughly the same distance—and height.

To jump an animal must do work. The technical definition of work is a force multiplied by the distance through which the force acts. Muscle provides the force. Big muscles are associated with strength; in fact, the force that can be exerted is proportional to the cross-sectional area of the muscle. An area, of course, is a length times a length. The work the squirrel puts into his jump, force times distance, is, therefore, proportional to length times length (from the area of muscle) multiplied by another length (the distance through which the muscle force acts); all lengths relate to dimensions of the animal.

In the case of a squirrel jumping straight up—up is easier to analyze than other directions and the conclusion is the same—the squirrel's work (work input) is used to lift himself a certain height (work output). The work output can be expressed as the weight of the squirrel multiplied by the height he lifts himself. Weight is proportional to size, or volume (length times length times length). Our equation, work input equals work output, therefore, looks like this: The left side of the equation, force times distance, has length times length times length; the right side, weight times height jumped, has length times length times length times height. On both sides of the equation we have length times length times length and can cancel them. What remains is an equation that states that the height an animal can jump is independent of the dimensions of the animal—all the dimensions have been canceled out of the equation.

This does not mean that the height jumped will be *exactly* the same for animals whose weights differ enormously. It does mean that the height will vary little compared to the variation in weight. The variation that does exist comes about because some animals have legs that are shorter or longer relative to their body size, or have legs that are attached to the body in a different way. A horse may weigh thirty times as much as a dog, for example, but they both clear roughly the same height. (A flea's jump may be different because air resistance plays a role with the very small. An elephant is at the other extreme: It can't get off the ground.) What the height jumped *does* depend upon is the density of the animal, the force the animal can exert for a given muscle area—and both of these are roughly the same for all animals—and the "strength" of gravity. All animals—and humans—can jump six times higher on the moon than on Earth because the pull of gravity on the moon is one-sixth that on Earth. What was impressive in the squirrels' leap to the fire-escape ladder, then, was not the distance but the shrewdness and the agility in the ricochet off the wall.

Blackbuster was back a day later in the afternoon. He had not altered his opinion of, or his ability to eat, any of the victuals I had offered the day before. I had some new ideas, however. I remembered a scene Larry had once described: deserted chess tables in

Washington Square Park on a cold winter day, a squirrel squatting on top of one of them, a partly eaten apple, bigger than his head, held in both arms. I took an apple from the refrigerator and cut a slice. Blackie was able to scrape it. He consumed the entire slice, leaving the skin. Meanwhile, I had acted on another remembrance of things past—Runty. Blackie appeared to be in as poor condition as Runty had been in once, and I thought milk might be in order. (Someday I'll try chicken soup.) It was inspiration on my part to serve the milk in the plastic cup. After a second slice of apple—*sans* peel—Blackbuster thrust his nose into the cup for investigatory purposes. He pulled back in shock. His face was wet. He licked his lips. He evidently decided that milk wasn't bad, for he thrust his head back in for a long drink. I could hear him lapping. He finished another slice of apple before leaving. I was thrilled. He had had about a quarter of a medium-sized apple and about an ounce of milk.

The following afternoon, I was even better prepared—with peanut butter. Blackbuster had apple to start, a swig of milk, and a peanut butter sandwich on pumpernickel. He loved the peanut butter. He did not love the pumpernickel. He left it, after removing the peanut butter. Making greatly exaggerated mouth motions to deal with the peanut butter, he swallowed a chunk and—horrors!—it stuck in his throat! He closed his eyes halfway and ceased all movement. He remained motionless for at least thirty seconds. I thought he would die. Then he opened his eyes to their full width, and took another mouthful of peanut butter.

I took pictures of Blackie as he drank his milk. He kept his eyes above the rim of the cup, fixed on me—God knows what I might do to him if he wasn't looking—his hind legs stretched far back in cautionary position. Before he left, he took a stab at a piece of Brazil nut and a half-walnut. He managed to make a small dent in each.

The next day, he tackled a pecan with most of its shell

removed. He polished it off in half an hour. He was ready for a rest then, in the form of a peanut butter sandwich. Then followed three quarters of a walnut—another half-hour—more peanut butter, a slice of apple, a drink of milk, another slice of apple, which he didn't finish. He was stuffed. He left in infinitely better shape than four days earlier.

The next three days, Blackie ate only nuts. No more milk and peanut butter sandwiches. For a few moments, I thought it was not Blackie. His coat looked sleeker, but there were the familiar scars on his face. The last remnants of winter fur must have been shed. The last of the three days, he arrived with a new wound, an open, bloody sore above his left eye. What was going on? Was this sweet, gentle Jekyll also an agressive Hyde? The wound over his eye did not interfere with his eating.

Squirrels heal rapidly—I suppose they have to, to survive in the wild. From death's door to filberts in a week.

Four days later, I guess I didn't hear Blackie when he arrived at the window, so he tore off the screen. He was well.

20. RULES AND GENERALIZATIONS

IF I WERE ASKED how many nuts a squirrel eats, I would reply with all sorts of qualifying statements. Genius, coming regularly, ate six or seven filberts each day. Holey ate three times as many, but Holey didn't come every day. He might have been somewhat of a lion, loading up for a three-day fast, in which case his average daily consumption was about the same as Genius's, or he might have sandwiched in other restaurants between visits to ours. The biggest eaters of all were some of the females, particularly when they were pregnant. Dainty little Ninotchka weighed in as champion with thirty nuts at a single sitting.

Squirrels are not like dogs or humans; they will not eat everything put before them. They are more like cats: They eat only what they need. One does not see fat squirrels (although one does hear of fat cats). When squirrels are well fed, they may be plump, but never obese. Humans and dogs are gourmands, cats and squirrels gourmets.

It might seem odd to call squirrels gourmets, when very nearly all they eat is nuts. But squirrels are connoisseurs of nuts, not unlike human connoisseurs of wines in discernment in a limited sphere.

That squirrels get all their nourishment from essentially one food I found pretty impressive—until I thought about cows. Cows get not only all *their* nourishment but they manufacture ours, too —from grass, which seems a lot less nourishing than nuts. What is impressive, then, is actually *nature.*

What's in a nut? My cookbook provided the information that twelve to fourteen almonds or three pecans or four to eight walnuts or twenty to twenty-four single peanuts contain one hundred calories. I estimated, then, that a typical squirrel meal—say, one Genius might eat—is, at most, one hundred calories. Genius weighed about one pound. If we assume that one meal was all he had for the day, and compare his calorie intake with that of a one hundred-fifty-pound human, we should multiply by one hundred fifty, the ratio of the two weights. The human would have to consume fifteen thousand calories per day to match Genius. A stevedore consumes fewer than five thousand calories. And, to compare with Ninotchka when pregnant—she ate four times as much as Genius—the human would have to consume sixty thousand calories. Can you imagine the offspring the human would produce from that kind of eating? A good many of the calories our squirrels consumed probably went into running up and down the fire escape to our apartment, and radiating heat through their large surface-to-volume ratio.

There were fashions in foods. All of our squirrels were excited over Brazil nuts at first. (Because they were large? Exotic?) Then Brazils went out of style; everyone left them for last.

Some tastes were pretty refined. Halfy, usually our most eclectic eater, when able to exercise choice in acorns, ate those that came from the refrigerator and not those I had left in my office at Rutgers for several weeks. It is possible that they were not the same kind of acorn.

Even acorns have, in addition to a complex biology, a history and a sociology as well. One batch of acorns had been imported from Canada. The mother of one of my students, Andy Simonson, an animal lover, had responded to my classroom appeal by bringing bagfuls home from a trip; she was worried that the drought in New Jersey that year might cause problems. A second student's mother was delighted that by providing my squirrels acorns her lawn would be clean. That latter batch caused difficulties for me at the end of the term when grades had to be assigned. The student was on the border between a B and a C. If I were to give her the B, was I being bribed? By acorns?

Larry asked me how the acorns were selling. I replied that our squirrels appeared to prefer nuts.

"What do squirrels in forests eat?" he asked.

"Acorns—when they can't get nuts," I replied.

"That means that squirrels are always eating what they don't like," Larry mused.

"Not what they don't like," I amended, "but what they like second best. What about all the people who eat rice when they'd rather have steak?"

I remember walking in Central Park one winter when I was in high school, and seeing a squirrel look at me with entreaty. All I had that was in any way edible was a cough drop. I offered it to the little fellow. He not only took it, he ate it.

Which nuts get eaten, and which are buried? That, too, varied, *Chaque* squirrel *à son goût*. Genius, in his prime, threw peanuts down with contempt. Yet peanuts were important as "fast food" for squirrels in a hurry—females, for example—or squirrels with dental problems. Ninotchka ate peanuts only when she was pregnant. That *could* mean peanuts are to squirrels what pickles are to humans during pregnancy. More likely, it means that Ninotchka's teeth were in bad shape from nest building. Genius would bury almonds in the apartment and immediately consume filberts. Surplus filberts he buried, of course, both inside and outside the apartment. Other squirrels, Halfy included, would not normally bury nuts as important as filberts in the house. Only unimportant nuts, those with soft shells, were entrusted to indoor sites. Walnuts? Not once did I ever find one in the apartment. Such valuables were probably placed in safe-deposit boxes.

There does not seem to be a season for burying. Everything I put out is always taken, except in extremely cold weather, when the rule appears to be "eat and leave, before you freeze." Or perhaps the ground is simply too hard for burying. Crowds modify burying. When the gang's all here it is "Devour ye nutmeats while ye may, for someone's right behind you." Burying begins early in life, from what I can tell. (Fighting, too, begins early. I have a vivid recollection of two virtual *babies* wrestling. The recollection is so vivid because it was on Armistice Day.)

The order in which the different nuts are eaten might provide clues to the personalities of the different individuals in the same way that such things provide insight into humans—if we know how to interpret them. When I put a Jewish-mother meal on the windowsill—acorns near the entrance; peanuts a bit further down the line, then almonds, half-walnuts, cracked pecans or Brazil nuts with shells partly removed, filberts in the cup—some squirrels ate first and then buried. Others buried first and ate later. It was not a question of being hungry or not; this was in the morning, when nobody had eaten since the day before. And the "eat laters" *ate* later, when "later" was sometimes just a few minutes later. There was some assessment, in squirrel terms, of how many competitors were around, how long the screen would remain open, or how long the supply would last, and whether it was better to get the nuts into the stomach or into the ground. And it was *consistent.* The "eat firsts" nearly always ate first.

I couldn't help thinking about the contrast between me and Larry. He eats first what he likes best, goes on to next best and finishes that, then tackles what he likes least. He claims that, in any other order, lightning might strike him before he got to the best and he would miss out. I eat meat, potato, and vegetable all at the same time. Some of the squirrels ate more or less the way I do, when possible: a filbert, then an almond, then an acorn or peanut, then another filbert. Others ate more like Larry: only the nut they liked best. Some of the variation arose from sources other than simple preference. One source was the state of an individual's teeth. During the shelling of a filbert, for example, the sheller has to stop from time to time to click his or her teeth for a few seconds before resuming shelling. Although this clicking sounds much the same as anger clicking, it obviously involves honing of the teeth, in analogy to a knife being sharpened during the carving of a roast. (Perhaps the anger clicking not only sounds the same but *is* the same as the shelling clicking—with the teeth being honed for an object other than a nut.) Alternating hard-shelled with soft-shelled nuts gave the teeth a rest.

There didn't seem to be much variety in the actual mechanics of dining, other than the two filbert-shelling styles. I suppose those might correspond to the two most common methods, among humans, of eating corn on the cob: typewriter style, left to right, then back to the left for the next row; and the helical style, starting at one end, rotating the cob to remove all the kernels in a ring, then moving over to the adjacent ring, and so on, until the end. I have never known anyone to combine the two methods, or, for that matter, to vary the starting end. (I must investigate whether lefthanded persons or Israelis start on the right side.)

Some squirrels appear never to bury. They eat their fill, maybe take a filbert for the road, and that is it. This group consisted almost exclusively of older males.

I wondered about those nonburying males. Why didn't they? Were their womenfolk getting their meals for them? Were they snatching from the stores of others? Had they built up an enormous supply of nuts in their youth? (Not likely; the nuts would not have lasted that long.) Did they have intimations of their mortality? Had they decided they wouldn't need nuts for a distant future?

Larry and I, too, had different shelling styles. Larry's pecans—for squirrel consumption—could be recognized by the small rectangular window in the shell. My pecans have a thin crack, which I widen, as desired, with my fingers. (I learned to like pecans, under the influence of the squirrels.) I marvel at how we quickly develop skills, in all manner of things. In the beginning, my Brazil nuts would splatter in the nutcracker. Now, I can remove only the curved surface opposite the acute angle, leaving the rest for use as a plate for a squirrel. My walnuts are two neat halves. I learned exactly the right amount of pressure to apply, and in a relatively short time.

Everyone scraped or chomped almonds on the shells a few times before transporting them out. If this practice is related to the rolling of a filbert to test whether there is a worthwhile nut inside, I cannot see how, unless the chomp punctures the shell and facilitates assessment through odor.

While some may be fastidious—the Matriarch, for example, who shelled already shelled nuts—on the whole, squirrels are sloppy eaters. Crumbs abound, particularly from almonds. Cookies are handled even more sloppily, and in addition I disapprove of junk food for squirrels, so I rarely dispensed them. Still, nature takes a wider view of such things: We acquired our own little flock of sparrow "cleaning women," who flew over at opening time when they saw the squirrels. They had learned that there would be a meal of crumbs.

One practice I cannot understand is that of the single bite. It is expressed every year on the apple trees in our garden. Each tree produces thousands of small apples, about the size of large cherries, and each apple has one bite taken from it, of a size that would fit a squirrel's mouth. Why *one* bite? If the taste is so bad that one cannot tolerate more than one bite, why keep on trying more apples after the first hundred? Will nothing discourage a squirrel in a fruit tree? (Once a black squirrel with a cauliflower ear, whom we never got around to naming, must have picked a bad almond. He gagged, left half the almond, and ran out of the apartment, still gagging. The half-almond looked and smelled all right to me; I assumed, therefore, that the first half had stuck in his throat, and put the remainder outside with some other nuts. When our cauliflower-eared friend came back, he wouldn't touch it.) The single-bite syndrome apparently extends to pears and some flowers. The families of two of Larry's colleagues consist of rabid squirrel haters. Norma Stroke reported two hundred pears felled in a few hours by a single bite near the stem of each, while Larry Bornstein testified that the very instant his prize tulips bloomed they were decapitated—not eaten—victims of wanton massacre.

The practice of each squirrel having a fixed eating spot sometimes made me think of my father and his special place at the table. Larry's father, too, had a special chair. And the couple of females who ate on the bed or the armchair when we were not in the room, and fled when we entered, reminded me of Larry's and my behavior when we sat in our fathers' chairs.

Females for whom the cup was "off-limits" when it was in its usual spot *would* take nuts from it when I moved it to a new location, such as the desk.

I am occasionally asked, when I comment on the lower status of female squirrels, whether there is "feminine behavior" in squirrels. Apart from some whining, which I have already reported, and which may say more about my prejudices than about squirrel behavior, there were only a couple of acts that I could point to as "feminine"—in human terms, that is. One was a female waiting on the fire escape, watching, while her man attempted to break into our apartment. The other was a female, having just been chased by a male, "hiding" behind one of the thin vertical bars of the fire escape, completely visible, of course. Perhaps it was not meant to be taken as a serious hiding place.

Some mornings, the breakfast scene in our bedroom was a model of zoodomesticity: Larry with his tea and I with my coffee, both on the bed; Sweetie on the desk with his walnut bits, Halfy under the desk with anything she could get hold of; Curly Ears on the ledge atop the lower half of the window—uneasy if the blind was slanted so that she couldn't see us—with an almond; and Holey or some other male ambling across the windowsill to the cup of filberts. The scene sometimes extended out onto the fire escape to the little sparrow maids, the blue jay flying off with half a walnut, and further out to the two pigeon camp followers watching from the roof of the next building, awaiting their turn to inspect the shells on the fire escape.

Larry is a major fruit eater. He believes tangerines are proof of the existence of God. He is also a major cheese eater. When dining on one of my "no cooking" meals (which he evidently prefers to my cooked meals), after consuming grapes, tangerines, brie, goat cheese, a mango perhaps, he is given to repeated exclamations of "Nero never ate like this!" One morning, surveying the breakfast scene, which included canteloupe seeds I'd washed the night before for Halfy, Larry added, "And Nero's *squirrels* never ate like this!"

21. URBAN
WOES

IN THE SPRING OF 1981, there were rumblings of trouble. Half a dozen new squirrels started showing up, and the Mews house directly in back of ours got a new tenant. Ordinarily, these would be two quite separate phenomena. But the new Mews tenant commissioned a lot of work on the house, and also decided to beautify the garden. Large, rectangular planters appeared. Then flowers appeared in the planters. Excess squirrels and new flowers are not a great combination. There was talk among the Mews tenants—who act as if the garden is theirs exclusively—of putting out poison. The new tenant appealed to Mr. Lassiter, and he revealed who it was that was feeding and abetting the squirrels. Mr. Lassiter must have decided that we should fight it out between us, for he also informed the new tenant that I would never stop the feeding. One Saturday morning, I got a visit from Mrs. New Tenant. She and I were very civilized and pleasant to each other—outwardly. She told me she did not want to deprive me of my "pets," but they were doing an enormous amount of burying, digging up, and reburying in the new flower beds she had installed —at her own expense—to beautify the garden for all. I did not express any of my immediate responses, namely, that none of the tenants in my building had either asked for or been consulted on the beautification, that the flowers were right near her house, and that the tenants in my building had been discouraged from using the garden by some of the Mews tenants. I said none of this because I knew that in any case I would have to do something about the squirrels. I was worried about Halfy, for one thing. (She disappeared right about this time.) Then we would soon be going away for most of the summer, leaving Jeannie with a population problem that would not cease at the end of the summer. (Larry and I were to have a one-semester sabbatical; we would be living in Princeton and would return to New York every few weeks. When I discussed the next few months and the rambunctious new squirrels with Jeannie, she had said that if I didn't get rid of a few of them, it would be "a lo-ng fall." I told the new tenant, therefore, that I would shell or crack all nuts, so that there would be none to bury; further, I would cut down the daily rations over the next few weeks.

I felt awful. It was the same business all over again that I'd experienced with the three marauders in Genius's time, only with more squirrels. Now I was to act like President Reagan, dispensing only to the "truly needy."

In the ensuing weeks, I cracked nuts and reduced rations, and several squirrels departed. One male with an extremely short tail resisted—he tugged and rattled the screen several mornings in a row—and the black female with one ear split like Halfy's managed to pull the screen out of its moorings right in front of her new offspring. That new offspring troubled me. The older squirrels would undoubtedly return to previous haunts. But that youngster had no previous haunts, and might starve unless his mother took him or her in hand. Also, now that there had been a demonstration of the pulling of the screen . . .

Before leaving at the start of summer, I reinforced the barricades on the fire-escape window. Outside the screen, which was outside the filter, which was outside the locked, metal-reinforced window, I placed an additional screen, one chewed and with several holes, which would serve probably only as a psychological deterrent. Jeannie would use the apartment from time to time—when hers was being painted and when she needed peace —but she would turn on the air conditioner rather than take down the barricades to open the fire-escape window, and so the squirrels would not suspect that anyone was within. I hoped for the best.

The best was pretty good, all summer. Jeannie was very busy and simply put nuts out on her windowsill; she was less knowledgeable than on any other occasion about who the recipients were. But the neighbors were not complaining, there did not appear to be too many squirrels around, and those that were around were not complaining, either. Jeannie had not seen Halfy all summer.

Jeannie continued on squirrel duty when we went to Princeton, with an occasional assist from Mrs. Russell, who was no longer working—she had reached ninety-five—and therefore had more time. October passed and all was well.

In early November, at quarter past eleven, after an evening out at the home of some friends in Princeton, we got a phone call from Jeannie. There had been a break-in at our apartment—by a four-footed housebreaker. Jeannie surmised that the intruder had shinnied up the space between the upper and lower halves of the air-conditioner window and chewed through the foam rubber insulating strip between them. A great deal of time had apparently been spent in the apartment, as there was incriminating evidence in every room of activities ranging from minor mischief to major mayhem. Jeannie thought the housebreaker had blasted out through the plastic partition between the air conditioner and the window frame. As a temporary measure, she bricked up that escape route with the bricks I keep under the radiator in case of need (in reality, because I am unable to part with anything). I told

Jeannie where she might find some wood with which to temporarily plug up the point of entry.

Two days later, I went to New York to assess and repair damage. In the living room, the Papago Indian basket and the nutcracker, normally on the chair-ladder, had been knocked to the floor, as had two tin Mexican candlesticks that had stood on the mantelpiece and the two copper ladles that hang from bricks on the fireplace. The fireplace screen was overturned, piles of soot dislodged from it onto the rug. In the bathroom, the spider plant had been knocked down from the windowsill and its pot broken. Soil was all over the floor. Jeannie had rescued the plant and put it in a new pot. The kitchen was untouched except for the rubber seal on the lower left corner of the refrigerator door, which had been chewed but not chewed through. The main damage was in the bedroom and pointed to a major effort to get out. On both windows, the central strip of wood, which divides the lower half of the window into two panes, had been chewed almost to the glass. And the upper border of each lower-half window—but mostly the air-conditioner window —was chewed in such a way that the space between upper and lower halves, normally less than one inch, was now two inches wide. Two pieces of wood, each approximately one inch by forty inches, had, to a depth of about half an inch, been rendered into sawdust, sawdust that covered the sill, the bookshelves, and the floor.

I concluded that Jeannie had been wrong about where the perpetrator (or, as we heard a member of the New York City Police Department say in a radio broadcast, "the perp"—pronounced a bit more like "poip") had entered. A squirrel could not have squeezed *up* between the two sections of glass—*down*, possibly, with difficulty, but not up, not with nothing to grasp. *Entry*, not exit, had been through the air-conditioner partition. (When we left New York, two weeks before the break-in, it had passed through my mind that that section might be vulnerable, but that thought had been banished by a reassuring one, that the squirrels probably did not connect that window with the one that supplied them with nuts.) The sequence of events was, most likely, entry through the plastic partition, exploration, discovery that there were no nuts to be had, then, attempts to get out. The point of entry had, by then, been forgotten or not recognized, since, from inside, it was hidden behind the nephthytis plant.

The quantity of sawdust and wood shavings was formidable. It bespoke hours of effort to escape. The tipped-over African sculpture on the headboard and the pawprints on the desk and armchair testified to wide ranging explorations in the bedroom. Soil on the file cabinet told of a search in the asparagus fern.

One thing puzzled me. I thought I knew who the perpetrator was even before receiving confirming evidence: the baby black squirrel whose mother, the "shy" one with the split ear like Halfy's, had demonstrated that when the nuts one wanted were not forthcoming, one pulled the screen out of its moorings. The confirming evidence was the appearance of said squirrel on the air conditioner during my hammering a piece of wood over the torn plastic partition. I had foolishly given that squirrel some nuts the last time we were in New York, and I was not about to repeat the foolishness now. The thing that puzzled me was the chewed corner of the refrigerator. That pointed to the little gray, Boldy, who would disgustedly walk into the kitchen if the cup was empty and neither Larry nor I showed signs of getting up from our places in the bedroom to accommodate him. But Boldy was not the type to break in. He was a respecter of property. What must have happened, since the little black had never been encouraged to enter the apartment, was that Boldy had followed the other onto the scene and there were two perpetrators. Otherwise—was it possible that nuts in the refrigerator could be detected from outside by the nose of an expert?

When we returned for good, in the middle of January, we kept discovering bits of damage we hadn't noticed at first, such as the toothmarks on the wooden screw-type nutcracker that had been knocked down in the living room.

No squirrels appeared for several days after we returned. We had driven home in a blizzard, and no visitors could be expected with a great deal of snow on the ground. I left plenty of nuts out anyway, to make the icy trip up to the apartment worth it to anyone who tried. The only squirrel who finally showed up was the one with the very short tail whom I had tried to get rid of in the spring. The fact that he was the only one around, plus his biting the window frame and rattling the filter a couple of times, made me shift suspicion regarding our housebreaker away from the little black. I was not overjoyed having a delinquent partaking of my largesse. However, when several weeks went by and no other squirrel put in an appearance, my attitude toward him changed. I began to bestow personal attention upon him at mealtime and I gave him a name. Quartertail was rejected in favor of the more euphonious Minitail. What continued to bother me was that we were down to one-third of a squirrel; only Mini was coming, and then only every third day.

Mystery surrounded Mini. Apart from the cloud of suspicion he was under as a possible housebreaker, and where it was that he spent the two days between visits to us, had he played some role in the disappearance of the other squirrels? And where was the rest of his tail? To compound the mystery, one morning Mini appeared with a strange "skin disease." Fur was missing and there was a kind of encrustment between his shoulders. The malady lingered for a couple of weeks, during which time some fur grew back and the encrustment changed in appearance. Sometimes it looked like a dirty strip of adhesive tape. Mini would not permit close examination of his affliction. Then, one day, he had wings. It was weird. Through the filter, he looked like an angel squirrel—this possible perpetrator. I hurriedly slid aside the filter. The wings were made of Kleenex. The "encrustment" was gone—it had indeed been a strip of adhesive tape—and the tissue had been substituted. Underneath, Mini had a wound, part open, part scab. Someone was ministering to Mini. Who, I still do not know.

We discussed the dearth of squirrels with Engelbert Schucking, a colleague of Larry's at New York University, who had been visited for the last couple of years, in much the same way we were, by a squirrel he named Tensing. What was special about Tensing, and the reason for his name, was that Schucking lived in an apartment house on the west side of Washington Square Park on the *fifteenth floor*! After eating on the windowsill, Tensing would sometimes climb up the brick wall to the roof, one floor above, and sometimes he would go down, dropping from air conditioner to air conditioner—some separated by four stories—using his tail as a parachute-rudder. And I was upset when *our* squirrels jumped from the fire-escape railing to the air conditioner, a horizontal distance of only five feet, and a mere five stories up! Tensing may live in the magnificent ten-story English elm in the northwest corner of the park, which some claim is the oldest tree in the city, and under which the turtle Larry brought to our marriage is buried. Anyway, Tensing was visiting as usual, and even bringing a friend on occasion.

Once again, I called the Museum of Natural History. I wanted to know if there had been a "squirrel blight." (Years ago, in Aspen, when for weeks the City Market had been out of the small boxes of raisins we liked to carry on hikes, we asked a girl at the checkout counter what the problem was. She answered, in complete earnest, "I think there's been a raisin blight.") I had a long, pleasant conversation with a Ms. Lawrence in the mammal department. I asked whether anything had occurred to reduce the squirrel population in the city. Had the winter been especially hard or the food supply low? Ms. Lawrence said that the squirrel population had increased near the Peter Cooper apartments and near the Williamsburg Bridge, and the squirrels were sleek and looked in fine shape. She ventured to guess that the squirrels may have decreased in Washington Square Park because of the density of the human population. "The squirrels may have found that Washington Square Park is not a nice place to live in anymore." It was true that the weekend bongo drums and steel bands were driving our friends crazy in the New York University apartments on the west side of the square. Squirrels might also be driven crazy. "Would they move," I asked, "through city streets?" "They've been known to do just that," she replied. I had visions of dark little figures setting off into the sunset, carrying sticks with bundles tied to the ends. The large tree I had looked upon as an apartment house for squirrels had been removed—whether as a victim of disease or of a hurricane, I no longer remember—in a kind of urban renewal. "Could they have died?" I asked with some hesitation. "You would have seen carcasses," Ms. Lawrence replied. I said nothing, but I thought of the section in Lewis Thomas's *The Lives of a Cell* in which he said he had never seen a dead squirrel. The only one *I'd* ever seen, other than on a road, was the black one that Cosmo, the handyman, had in a box; it had died trying to get out of a drainpipe on one of the Mews houses. I had gone to Mr. Lassiter's office to identify the body, fearing it might be Blackbuster. It wasn't.

There are lots of squirrel "artifacts" around the apartment— a drawing of a squirrel, a blue enamel-on-silver ashtray with a squirrel on the side, several small ceramic squirrels—all gifts from people who know about our squirrels.

A while ago, I was awakened by a distant three-way argument involving a squirrel, a blue jay, and a cat: first the *waah, waah,* then the jay's raucous note, then a *meow,* the same sequence over and over until the *waah waah* changed to a *bok bok bok bok bok.* People always say Manhattan is a jungle.

We are still host to our one-third of a squirrel. And, as the squirrel is male, we don't expect to see any progeny on our sill. His two-day absences may even indicate he's not a local but a commuter (or, more likely, that we're not his only sponsor). Reflecting on the last dozen years, I feel that I have been blessed.

I once read an essay by philosopher C.E.M. Joad in which he declared that pains were additive: One could have a pain here, another there, and they would combine, symphonically, unlike pleasures, which detract from one another. Nevertheless, there are certain pleasures that add to the pleasure in life. They don't subtract from or take away anything; they simply add. Love of animals is one such pleasure. It doesn't detract from work or relations with family and friends; it is sheer bonus. I love dogs for their zest for life, their finding every blade of grass important, every hole worthy of an investigation. I love their responsiveness, their joy at seeing me. Others might like the power they have over a living creature. (Larry claims the difference between dog lovers and cat lovers is that dog lovers enjoy the power of saying "Sit!" and being obeyed.) Not me. Creatures—all—have power over me. When I rode horses, they sensed immediately that I could never touch them with a twig, let alone kick them. They would ignore my entreaties and stop for lunch, long drinks. Murderous ones tried to rub me off—and out—on trees, or to decapitate me on low-lying branches. I had no power over them, I didn't blame them. Why *shouldn't* they try to get rid of this burden on their back if they could? I spoil all animals rotten. The cats I like are like dogs: strong, sturdy, tough, *interacters*—not finicky and aloof, "independent"—cats you can wrestle with, like Kalmbach.

Some of the best things in life are indeed free. The pleasure to be derived from a gust of wind on a sunny spring day, the sight of a golden retriever looking back to see if you are following on the trail, the sight of the muscles on the shoulders of a sturdy little rodent, all are God's gifts—joy to mankind. And the people who can experience these joys are blessed.

EPILOGUE

MINI HAS A LADYFRIEND. I have named her Maxi, short for Maxitail, as hers is rather long. She is a complex character. She is very grabby, but her grabbiness stems from fright: She shuts her eyes as she grabs and simultaneously covers them by bringing her tail up over her head to form long bangs. Frightened as she may be, however, she has demolished the screen; she cannot—or refuses to—understand that the number of nuts that issue from our house is not infinite. At the moment, she is the most troublesome squirrel we have encountered. Nevertheless, I welcome her. The dynasty will continue.

There is also a new black squirrel.